T0205499

Nanotechnology

Muhammad Bilal Tahir · Muhammad Rafique ·
Muhammad Sagir

Editors

Nanotechnology

Trends and Future Applications

 Springer

Editors
Muhammad Bilal Tahir
Department of Physics
Khwaja Fareed University of Engineering
and Information Technology
Rahim Yar Khan, Pakistan

Muhammad Rafique
Department of Physics
University of Sahiwal
Sahiwal, Pakistan

Muhammad Sagir
Department of Chemical Engineering
Khwaja Fareed University of Engineering
and Information Technology
Rahim Yar Khan, Pakistan

ISBN 978-981-15-9439-7   ISBN 978-981-15-9437-3   (eBook)
https://doi.org/10.1007/978-981-15-9437-3

This Springer imprint is published by the registered company Springer Nature Singapore Pte Ltd.
The registered company address is: 152 Beach Road, #21-01/04 Gateway East, Singapore 189721,
Singapore

Contents

Historical Background, Development and Preparation of Nanomaterials

Umair Riaz, Tariq Mehmood, Shazia Iqbal, Muhammad Asad, Rashid Iqbal, Umair Nisar, and Muhammad Masood Akhtar

Abstract Nanotechnology has developed as a field of science with broad applications in different fields. The reduction of particle size and tuning the particle morphology of materials from micro- to nanosize leads to the unique properties and helps in versatile applications. The reason for the nanomaterials (NMs) to exhibit enhanced properties is due to the large surface-to-volume ratio and quantum confinement effect. It is therefore imperative to understand the history, background, physicochemical nature before its usage. In this chapter, the overall overview of history and development of nanoparticles are discussed in detail. People have learned the way to produce bread, fabric, wine and cheese since ancient times when the process of fermentation was crucial. Secrets of ancient nano-production in many instances simply passed from generation to generation, without going into the reasons why the materials and products obtained from them acquired their specific properties. The development and preparation of nanomaterials some pre-conditions required for the specific preparation of nano-products. Various methods are used for the preparation of nanomaterials like chemical vapor deposition, hydrothermal method, hydrolysis, chemical engineers (heterogeneous catalysis) and biological route, etc.

U. Riaz (✉)
Agriculture Department, Soil and Water Testing Laboratory for Research, Government of Punjab, Bahawalpur 63100, Pakistan

T. Mehmood
College of Environment, Hohai University, Nanjing 210098, China

S. Iqbal
Institute of Soil & Environmental Sciences, University of Agriculture, Faisalabad, Pakistan

M. Asad
LokSanjh Foundation Bahawalnagar, Bahawalnagar, Pakistan

R. Iqbal
Department of Agronomy, Faculty of Agriculture and Environmental Sciences, The Islamia University of Bahawalpur, Bahawalpur, Pakistan

U. Nisar · M. Masood Akhtar
Agriculture Department, Government of Punjab, Regional Agricultural Research Institute, Bahawalpur 63100, Pakistan

M. B. Tahir et al. (eds.), *Nanotechnology*,
https://doi.org/10.1007/978-981-15-9437-3_1

Keywords Nanomaterial preparation · Nano-products · Nanoparticles · Nanotechnology history

1 Introduction

The word nanotechnology defines a variety of technologies with widespread applications conducted on a nanometer scale as enabling technology in various industries. Nanotechnology involves the development and implementation of physical, chemical and biological systems at scales ranging from single atoms or molecules to about 100 nm, as well as the incorporation of the resulting nanostructures into larger systems. This "i" dot area alone can encompass 1 million nanoparticles. What is distinctive about nanoscale materials compared to the same materials of greater form is that they can become more chemically reactive and change their strength or other properties due to their comparatively larger surface-area-to-mass ratio. Also, the laws of classical physics give way to quantum effects below 50 nm, causing different optical, electrical and magnetic behaviors. For decades, nanoscale materials have been used in applications ranging from window glass and sunglasses to car bumpers and paints.

Now, however, the convergence of scientific disciplines (chemistry, biology, electronics, physics, engineering, etc.) leads to a multiplication of applications in materials manufacturing, computer chips, medical diagnosis and health care, energy, biotechnology, space exploration, safety and so on. Thus, nanotechnology is expected to have a significant impact on our economy and society over the next 10–15 years, growing in importance over the long term as further breakthroughs in science and technology are achieved. On the one hand, this convergence of science and increasing diversity of applications on the other is what drives the potential of nanotechnologies. Indeed, their greatest impact may arise from unexpected combinations of previously separate aspects, just as the internet and its myriad applications arose through the convergence of telephony and computing [30]. The catch-all term "nanotechnology" is so broad that it is ineffective as a guide to addressing issues of risk management, risk management and insurance. A more focused approach is expected concerning all related aspects of risk management. In terms of risks to health, the environment, and safety, almost all of the concerns raised. About health, environmental and safety risks, nearly all concerns raised relate to free nanoparticles, rather than fixed nanoparticles (www.oecd.org/chemicalsafety/nanosafety/44108334.pdf. Accessed 5-5-2020).

2 Background

Even long before the beginning of "nano-era," people came across various nanosized structures and the associated processes of nano-level and used them in action. Nevertheless, intuitive antiquities in nanotechnology evolved naturally, without proper comprehension of the essence of certain structures and processes. For example, the fact that small particles of different substances possessed different properties than those of the same substances with larger particle size was long known, but the explanation for this was not clear.

So, people were subconsciously engaged in nanotechnology, without guessing they were dealing with the phenomena of the nanoworld. Secrets of ancient nano-production in many instances simply passed from generation to generation, without going into the reasons why the materials and products obtained from them acquired their specific properties. People have mastered the ways of making bread, wine, beer, cheese and other foodstuffs since ancient times, where fermentation processes at the nano-level are important. In ancient Egypt, dyeing hair in black was very normal. It was assumed for a long time that the Egyptians mainly used natural vegetative dyes, henna and black hair dye. Nevertheless, recent work into hair samples from ancient burial sites in Egypt, performed by Ph. Walter, and he found that hair was dyed in black with lime paste, lead oxide and small quantities of water. Galenite nanoparticles (lead sulfide) were produced in the course of the dyeing process. Natural black hair color comes with a pigment called melanin which is spread in hair keratin in the form of inclusions. The Egyptians were able to react with sulfur, which is part of keratin, to the dyeing paste, and obtain a few nanometers of galenite particles in size, producing even and steady dyes. In the Middle Ages, the manufacturing development of multi-colored stained-glass church windows in Europe achieved great perfection. The windows, as recent work reveals, included gold additives and other metal nanoparticles. This bowl, on which Licurg, Edons' tsar is depicted, possesses unusual optical properties: It changes color by changing the position (inside or outside) of the source of light. In natural light, the bowl is white, and it turns red when illuminated from the inside. The analysis of bowl fragments, conducted in General Electric's Laboratories for the first time in 1959, showed that the bowl consists of the usual soda–lime quartz glass and has about 1% of gold and silver and also 0.5% of manganese as components. The researchers then believed that colloidal gold accounts for the unusual color and disseminating effect of glass. Later, as testing methods became more sophisticated, scientists were using an electronic microscope and roentgenograms to discover gold and silver particles from 50 to 100 nm in size. It is those particles that are responsible for the bowl's unusual coloring. The European knights faced for the first time in the days of crusades the incredible power of blades in battles against Muslims, made of an ultra-strong Damask material. Any attempt to recreate such steel by armoring medieval Europe had no success. Using an electronic microscope, Paufler researched saber fragments made from Damask steel. The tests showed that the steel's structure was nanofibrous. Such a structure is supposed to have been obtained after a special thermomechanical steel processing made from the material of a different structures [30].

3 History

People have learned the ways of producing bread, wine, beer, cheese, and other foodstuffs since ancient times, where the processes of fermentation at nano-level are crucial. Hundreds of years, BC people knew natural fabrics and used them: flax, cotton, wool, silk. They managed to grow them and process them into products. The fact that they have an established network of pores with the size of 1–20 nm, i.e., they are traditional nano-porous materials, makes these fabrics unique. Natural fabrics possess high utilitarian properties because of their nano-porous structure: They absorb sweat well, swell fast and dry [30]. Taniguchi first brought the term "nanotechnology" into a scientific world at 1974 International Conference on Industrial Manufacturing in Tokyo to explain the super-thin processing of nanometer accurate materials and the development of nano-sized mechanisms. Nanotechnology strategy ideas that Feynman put forward were developed by E. Drexler published in 1986 in his book "Vehicles of Creation: The Advent of the Nanotechnology Era." A number of important discoveries and inventions were made during the second half of the 1980s to the early 1990s, which created a significant influence on the further development of nanotechnology.

In 1991, National Scientific Fund's first nanotechnology system started running in the USA. In 2001, the United States National Nanotechnology Initiative (NNI) was formulated with the idea of this program as follows: "The National Nanotechnology Initiative defines the interaction strategy between the federal departments of the United States intending to prioritize the development of nanotechnology, which in the first half of the year should become the basis for the economy and national security of the United States".

A special committee of the American Center for Global Technology Assessment monitored and analyzed the development of nanotechnology in all countries during 1996–98, prior to NNI approval, and published the survey newsletters on the basic development trends and achievements for US scientific, technical and administrative experts. The interbranch community session on nanoscience, nanoengineering and nanotechnology (IWGN) was held in 1999, the outcome of which was the nanotechnology work forecast for the next 10 years.

In 2000, under the auspices of the Industrial and Technical Committee, the Japanese Economic Association organized a special department on nanotechnology, and in 2001, the Framework Plan for Nanotechnology research was developed. The main provisions of this plan were: to designate the following as the basic "breakthrough" directions of nanoscience, information technology, biotechnology, power, environmental studies and materials technology; to make substantial investments in nanotechnology-based manufacturing; to intensify research in the above-mentioned areas and to apply their results to manufacturing in a manner that would allow for significant capital investment in nanotechnology-based manufacture; intensify research in the above-mentioned areas and apply their results to manufacturing so that they become "drivers" of the future revolution in nanotechnology; develop

a national strategy for the development of nanotechnology and, above all, organize effective cooperation between industrial, state and scientific departments and
research organizations.

Western European countries perform nanotechnology research within the context
of national programs. In England, the development of nanotechnology is supervised
by the Science and Technology Research Council and the National Physical Laboratory. In France, the National Center for Scientific Research defines the nanotechnology development strategy. In Germany, nanotechnology research is mainly funded
by the Ministry of Education, Science, Research, and Technology. Production of
nanotechnology in China, South Korea and other developing countries is getting
more and more coverage. Nanotechnology research has recently started in the CIS
countries, typically within the context of state science programs. Thus, at the turn
of the 1960s, the nanotechnology paradigm was created, while the 1980s and 1990s
are the beginning of nanotechnology growth in its own right.

Accordingly, one might argue that the entire time up to the 1950s can be regarded
as the nanotechnology prehistory. The end of this era was the emergence of conditions for controlled development of nanotechnology, facilitated by the scientific and
technological revolution which marked the second half of the nineteenth and the
beginning of the twentieth centuries [30].

4 Natural Nanomaterials

In many areas of our everyday life, innovative materials with diameters from a range
of one to 100 nm have also been found to have the impression that those materials
are "not entirely natural." The rapidly evolving field of nanotechnology has created a
wide variety of nanomaterials over the last few decades, most of which are considered
"synthetic" and are often met with some skepticism in the general public [44].

The physicochemical characteristics of synthetic nanomaterials are remarkable,
adapting them for a wide variety of medical, pharmaceutical, biotechnological and
energy applications. Synthetic nanomaterials include (1) carbon-based nanomaterials such as carbon nanotubes (CNTs), graphene and fullerenes (C60 and C70); (2)
quantum dots (such as CdSe and CdTe); (3) metal-based nanomaterials including
zero-valent metals (such as Au, Ag and Fe ENMs), metal oxides (such as nano-
ZnO, $-$ TiO2 and $-CeO_2$) and metal salts (such as nano-silicates and ceramics); (4)
quantum dots (such as CdSe and CdTe) (Hetami et al. 2016).

Synthetic nanomaterials may enter plant cells and interact with intracellular structures and metabolic pathways that can cause toxicity or promote plant production and
growth through various mechanisms, due to their ultra-finished size and high surface
reactivity [45]. The rapidly increasing use of synthetic nanomaterials has raised ques-
tions about possible impacts on live organisms and non-living (abiotic) ecosystem
components [29]. There is therefore little surprise about the field of nano-toxicology,
which was recently attracted to a particular degree, and concern about potentially
toxic effects on humans and environmental pollution with nanomaterials have also

increased [31]. This concern is not completely unwarranted, as some dramatic examples demonstrating the exhaust gases, smokes, and smoke fraction of the PM2,5 for example asbestos (mean diameter ranging from three to cinq microns) and other, airborne particulate matter [32, 33]. Produced nanoparticles displayed toxic properties in initial experiments. They can penetrate the human body in various ways, touch vital organs via blood and damage the tissues probably. The characteristics of nanoparticles differ not only from masses of the same composition, but they also show different patterns of interactions with the human body because of their small size. Therefore, a risk assessment is not appropriate for bulk material to classify the same nanoparticulate content (www.oecd.org/chemicalsafety/nanosafety/44108334. pdf. Accessed 5-5-2020).

While many of these materials are technically sound and sophisticated, nano is not necessarily "artificial." Nature itself is an excellent nanotechnologist. It supplies a range of fine particulates: inorganic ash, soot, sulfur and mineral particulates found in the air or wells and many bacteria and yeasts produce sulfur and selenium nanoparticles. These nanomaterials are entirely natural, and the development of natural nano-products is not surprisingly increasingly concerned. A number of polydisperse micro- and nanoparticles are found in the volcanic ash clouds. Such particles are between 100 and 200 nm in size and consist mostly of silicate and iron compounds. Severe respiratory disorders can occur during inhalation. Indeed, while particles of sizes within the lower micrometric range in the upper respiratory tract are deposited in depth beyond the nanometer range, they can cause serious respiratory disorders and deposition in tracheobronchial and alveolar regions [34]. The soot collected for "carbon nanotubes" from, for example, the combustion of Texas Pi'non Pine has several carbon nanotubes of a size of 15–70 nm. The carbon-based objects are readily airborne, posing significant health threats to animals and people [35]. Nanoparticles formed by natural nanosizing, therefore, are not uncommon in our environment. Fine platinum particles from millions of cars and their catalysts are also essential examples, and tires have abrasions which are difficulty "normal" but gradually but constantly affect the atmosphere [42, 43].

Both polydiverse nanoscopic and microscopic solid materials are in our drinking water. This is irregular in type and chemical-dependent primarily on $CaCO_3$ and $CaSO_4$, frequently mixed with iron oxides. Certainly, these particles are of very bad quality and cannot be compared to nanomaterials of modern nanotechnological processes that are perfectly formed, well-defined and homogeneous [37]. But, the fire in the inorganic sphere, which can eventually lead to nanoscopic particles, is just one chemical phase. Precipitation, oxidation, and, to a lesser degree, reduction often render inorganic materials into natural nanoparticles.

For example, abrasion results in the scrapping, cutting or grinding of large lumps in a fine particle material. Many such examples have been produced of naturally occurring nanomaterials such as $CaSO_4$ and spring water silicate particles [36]. Significant small particles are produced by chemical and physical processes such as weathering, slow rainfall by iron oxide particles, but also by carbonate dissolution and precipitation under CO_2 influence and hydrocarbon formation (HCO_3^-) intermediate [37]. In addition, inorganic matter such as refined nanoparticles of Fe_3O_4 and MnO_2 has recently inspired their colleagues to synthesize a wide variety of like particles based on natural materials [38], 39] as in the previous post. At the same time, Nature produces these "nanosize" pieces in our oceans, fish and food highlights sometimes with unwillingness in cahoots with human activities [40, 41]. Natural nanomaterials synthetic nanomaterials, and natural or naturally produced natural products nanomaterials, all have their own, special chemical and physical properties, their biological activities and their applications, especially in the field of medicine, food, cosmetics and agriculture. They are promising. In the future, these natural nanoparticles would not only promote innovation and offer a greener outlook for a conventional high-tech market, but they will also provide solutions with forgiveness for a variety of challenges.

5　Pre-conditions of Nanomaterials for the Development

Current regulatory structures are sufficiently wide and versatile to accommodate nanotechnologies, but UK and EU regulations tend to have a void in the identification of emerging chemical hazards, as they do not take particle size into account. All regulators must review the impact of nanotechnologies within their remit on the regulations. There is little after a substance has been approved on a bulk scale, to avoid the same substance being used as nanoparticles, which can be even more reactive or harmful without further research. The EC has already recognized the need to reconsider the mass thresholds which cause toxicity tests (www.europa.eu.int/comm/health/ph_risk/documents/ev_20040301_en.pdf. Accessed 9-5-2020). The US Environmental Protection Agency is also assessing whether to treat nanomaterials as new chemicals. Limits of workplace exposure to manufactured nanoparticles and nanotubes should be reviewed, as should procedures for managing accidental releases both inside and outside the workplace. For certain ingredients, including dyes and UV filters in cosmetics, such an assessment is already required within the EU [28].

Most nanotechnologies do not pose any new threats to human beings or the environment. At the nanoscale, materials will behave very differently to the way they do in bulk. This is both because the small particle size significantly increases the surface area and hence the reactivity and also because the quantum effects begin to become important. It means their toxicity in the form of larger particles may be different from that of the same chemicals [46, 47. Examples are where nanoparticles can have toxic effects even if the bulk substance is non-poisonous. This arises in part because they have increased surface area, and because the nanoparticles enter the body through inhalation, ingestion or absorption through the skin, they can move around and enter cells more easily than larger particles [28].

Few researches have been done on the toxicity of processed nanoparticles and nanotubes, but we can learn from studies on the effects of mineral dust exposure in certain workplaces and on the air quality nanoparticles. Important evidence from industrial exposure to mineral dusts indicates that the toxic threat is related to the surface area of the particles inhaled and their surface movement. It is quite doubtful that fresh, synthetic nanoparticles will be injected in sufficient doses into humans to induce the health effects associated with air pollution. Some can be inhaled in large quantities at some workplaces, and steps should be taken to reduce exposure. Newer particles with properties substantially different from these should be handled with special caution. Broad, thin asbestos-like fibers (less than about 3 μm and longer than about 15 μm) are a specific concern [48].

We have concerns about carbon nanotubes which, if inhaled in quantity as single fibers might cause similar problems to asbestos. Present processing methods tend to result in nanotubes clumped into "bundles." However, much of the current activity is aimed at developing techniques and coatings to keep the nanotubes separate. Because the nanotubes are designed to be insoluble, they may remain in the lung tissue and induce the inflammation free-radical release. The human exposure to airborne nanotubes in laboratories and workplaces should be minimized until further toxicological studies have been undertaken.

There is practically no information on the environmental impacts of nanoparticles. Factories and testing laboratories are advised to handle processed nanoparticles and nanotubes as if they are dangerous and aim to eliminate or remove them from waste streams. The industry will also determine the rate at which these components are released during the life cycle of the goods and materials comprising them and make this information available to the relevant regulatory authorities. Nanoparticles are envisaged to react with pollutants from the soil and groundwater to leave harmless compounds. Fe nanoparticles slurries have already been used [49]. It is recommended that the use of natural, processed nanoparticles in environmental applications such as cleaning up contaminated land should be banned until sufficient studies can prove that potential benefits outweigh the risks.

Where toxicological data is incomplete for ingredients integrating emerging technologies such as nanotechnologies, we recommend that the terms of reference of scientific advisory committees that consider their safety should require all applicable safety assessment data to be put in the public domain. Manufacturers should publish descriptions of the methodologies used in safety testing of any products containing nanoparticles to show how they took into account that their properties differ from those of larger forms [28].

6   Methods of Preparation (Overview)

Based on quantum mechanics principles, the material on the nanometer scale may have distinct physical properties [8]. It is generally acknowledged that the synthesis process plays a vital role in the advancement of nanotechnology applications and the characteristics of nanomaterials [11, 25]. The preparation methods provide an opportunity to induce desirable characteristics in a nonmaterial by tailoring the temperature, pH, mixing and agitation rate, base and gas concentration, and reactants ratio, depending upon preparation method [7]. Nanomaterials have versatile and vast range applications in daily life, including electronics, robotics, nanoscale optics, nano-biological system, and nanomedicine: Albeit, the application and functionality of nanomaterials substantially rely on the effective preparation of nanomaterials. Moreover, cost-effective and straightforward preparation approaches are vital for progressing in application and commercialization of nanotechnology. The preparation of nonmaterial highly acquires preparation ingenuity. Although it is easy to broaden a balanced preparation technique of nonmaterial [5, 9], there is continuously a chance of serendipity. A successful synthesis of a variety of nonmaterial has been carried out by an old-style ceramic method in the last several decades. The conventional ceramic method carried out by sieving, mixing and crushing the ingredient's powder (e.g., carbonates, oxides and other species), followed by heating and transitional grinding when required, respectively [7]. In modern nanomaterial preparation changing in extreme boundaries of influencers such as temperature, pressure, quenching and fugacities have been employed. Currently, researches are more emphasis on structural control, phasic purity and stoichiometry of reaction prevailed in nanomaterial preparation. The so-called soft chemistry strategies, which the French name Chimiedouce, are indeed vital as they result in new products, any of which are metastable and might not, in any other case, be synthesized. Such soft chemistry routes mainly employ simple reactions that are conducted at relatively low temperatures [7, 15]. To attain the specified traits of interest in nanomaterials, the method of preparation performs a critical role. Therefore, having expertise in synthesis techniques to achieve nanomaterial with the desired properties for a particular application are highly advantageous. Some of the main nonmaterial methods are chemical vapor deposition, hydrothermal method, hydrolysis, biological route, chemical engineers and so forth. Some of them were briefly defined inside the following sections.

7 Chemical Vapor Deposition (Brief Description)

The nonmaterial preparation via a chemical vapor deposition process was carried out by depositing the chemically produced solid material in gas or vapor form on heated surfaces [19]. The apparatus used to perform CVD comprises a deposition chamber equipped with a gas supply and exhaust system. The CVD process needs high activation energy, which can be given by various sources. For instance, the ultraviolet radiations, which are an excellent source of photon energy, are applied to break the chemical bandings in precursor molecules in photo-laser CVD. Hence, photon activation reaction can perform the deposition of material at room temperature [1]. Usually, the CVD method is used to obtain high-quality and efficient nanometer thickness thin films of solid materials with semiconductor properties [24]. The CVD process was previously used to synthesize nanocomposites of SiC/Si_3N at 1400 °C in the presence of various source gas such as H_2, WF_6, CH_4 and SiH_4 [13].

8 Hydrothermal Method

Hydrothermal synthesis is a solution reaction-based nonmaterial preparation technique. This process can produce nonmaterial at a wide range of temperature options starting from ambient temperature to very high. Hence, the conducive environment for synthesis of nonmaterial, which is usually unstable at higher temperatures, can be achieved. The morphology of desired nonmaterial can be controlled by regulating the pressure conditions depending on the vapor pressure of precursor molecules [6]. Several nanomaterials, including nickel ferrite (NiFe2O4), graphene sheets, quantum dots and titanium dioxide, have been prepared by using the hydrothermal method. Zhu et al. [27] obtained a more stable and high quantum yield of carbon quantum dots (CQDs). Similarly, Wang and Li [23] performed a hydrothermal method to produce hollow nanospheres titanium dioxide to perform catalysis and photoluminescence processes.

9 Hydrolysis

The hydrolysis or aqueous sol–gel method is a type of sol–gel nonmaterial synthesis approach. Hydrolysis nonmaterial preparation technique includes four steps, including hydrolysis followed by polycondensation than drying and finally thermal decomposition [16]. The oxygen required for this method to oxidize the metals for metal oxide formation is obtained from the water solvent. Mostly, metal precursors such as metal acetates, chlorides, nitrate, sulfates and metal alkoxides are used in this process. In metal oxide nanoparticle synthesis, metal alkoxides are more popular metal precursors due to their higher reaction affinity toward water [14]. This process has limited advantages for maintaining the homogeneous morphology of prepared nanoparticles [12]. It is a costly method compared to others and is mostly used to prepare maghemite (γ-Fe_2O_3) nanoparticles of size 6–15 nm [16].

10 Biological Route

Several studies successfully use biological routes to prepare non-materially. The biosynthesis of various metallic nanoparticles has been studied by using bacteria, yeast, fungi and virus [22]. For instance, biomineralization is a modern approach to preparing iron oxide nanoparticles. The two known modes of biomineralization are biologically induced biomineralization (BIM) and biosynthesis through Fe (III)-reducing bacteria (FRB). The BIM model is an extracellular preparation of magnetite, and it is highly influenced by external environments such as temperature, pCO_2, pO_2, redox potential and pH. The metabolites released by microbial cells generally from anaerobic bacteria react with particular ions or molecules which are adsorbed on the microbial cell surface in solution and mineralize them. The obtained minerals have poor crystalline structure [3]. Roh et al. [17, 18] reported *Geobacter* spp., *Shewanella* spp., and *Thermoanaerobacter ethanolicus* (Fe(III) reducing bacteria) while *Archaeoglobus fulgidus* and *Desulfuromonas acetoxidans* belongs to SRB are more common bacterial species which perform biomineralization. On the other hand, *FRB* is carried out by using *Shewanella putrifaciens* and Geobacter metallireducens that produce magnetite crystal as a byproduct during their metabolic activities [3]. In anaerobic conditions, bacteria obtain oxygen from Fe(III) oxyhydroxide and release poor crystals of Fe(II), which adsorb on ferric hydroxides beads and transformed into magnetite[3, 26]. These biosynthesis materials are economical, environmentally friendly and can be produced at neutral pH and room temperature. The famous example of biosynthesized iron nanoparticles is zero-valent iron nanoparticles [16].

Similarly, various other microorganisms are also being used to produce metal nanoparticles. Ahmad et al. [2] produced gold nanoparticles by intracellular biogenesis. They used actinomycete species, i.e., rhodococcus, and produced gold nanoparticles of 5–15 nm size range. Some yeast species, *Schizosaccharomyces pombe* and *Candida glabrata*, were reported to produce quantum crystallites in the presence of cadmium salt. The S. pombe strain showed excellently performed intracellular biogenesis of cadmium sulfide nanoparticles of 1–1.5 nm size with hexagonal lattice formation [10]; similarly, intracellular preparation of silver nanoparticles with 12–22 nm size was performed by using Haloarchaea [22].

11 Chemical Engineers (Heterogeneous Catalysis)

Metal nanoparticles have found as productive potential as heterogeneous catalysts. They can readily disperse over a surface and owing to large surface areas with plenty of reactive sites. They can perform significant catalytic reactions [21]. They are frequently used in the energy sector, chemical reactions and treating environmental pollution. Since many of them are thermally unstable and can only be performed in a narrow range of temperatures, it limited their usability in industries [4]. The surface modifier or dopant may produce a promising effect on the catalytic performance of nanomaterials by supporting the active sites against chemical reactions and high pressure attributed dynamic [20]. They explained that improving the morphology and electronic characteristics of supporting metal particles by doping of oxides material can significantly enhance the reactivity of nonmaterial. Moreover, atomic flexibility and small particle size of metal clusters can promote adsorption and catalytic characteristics of nanoparticles. Overall, the role of nanoparticles in herogeneous catalysis is not limited to as catalysis. However, they also have a very crucial contribution in sensor, photonics, biomedical imaging, energy conversion and storage and surface coating, etc., where virtually justify their stability and applicability in industrial processing [4].

References

1. Aeila, A.S.S., et al.: Nanoparticles—the future of drug delivery. J. Pharmaceutical Res. 9 (2019)
2. Ahmad, A., et al.: Extra-/intracellular biosynthesis of gold nanoparticles by an alkalotolerant fungus, *Trichothecium* sp. J. Biomed. Nanotechnol. **1**, 47–53 (2005)
3. Bazylinski, D.A., et al.: Modes of biomineralization of magnetite by microbes. Geomicrobiol. J. **24**, 465–475 (2007)
4. Cao, A., et al.: Stabilizing metal nanoparticles for heterogeneous catalysis. Phys. Chem. Chem. Phys. **12**, 13499–13510 (2010)
5. Chaturvedi, S., et al.: Applications of nano-catalyst in new era. J. Saudi Chem. Soc. **16**, 307–325 (2012)
6. Gan, Y.X., et al.: Hydrothermal Synthesis of Nanomaterials. Hindawi (2020)
7. Ganachari, S.V., et al.: Synthesis techniques for preparation of nanomaterials. In: Handbook of Ecomaterials. Springer, Cham (2017). https://doi.org/10.1007/978-3-319-48281-1_149-1
8. Initiative, N.N., What's so special about the nanoscale. Retrieved from Nanotech. 2018.
9. Koo, O.M., et al.: Role of nanotechnology in targeted drug delivery and imaging: a concise review. Nanomed. Nanotechnol. Biol. Med. **1**, 193–212 (2005)
10. Kowshik, M., et al.: Microbial synthesis of semiconductor CdS nanoparticles, their characterization, and their use in the fabrication of an ideal diode. Biotechnol. Bioeng. **78**, 583–588 (2002)
11. Mariotti, D., et al.: Plasma–liquid interactions at atmospheric pressure for nanomaterials synthesis and surface engineering. Plasma Processes Polym. **9**, 1074–1085 (2012)
12. Niederberger, M.: Nonaqueous sol–gel routes to metal oxide nanoparticles. Acc. Chem. Res. **40**, 793–800 (2007)
13. Rajput, N.: Methods of preparation of nanoparticles—a review. Int. J. Adv. Eng. Technol. **7**, 1806 (2015)
14. Rao, B.G., et al.: Novel approaches for preparation of nanoparticles. In: Nanostructures for Novel Therapy, pp. 1–36. Elsevier, Amsterdam (2017).
15. Rao, C., et al.: Soft chemical approaches to inorganic nanostructures. Pure Appl. Chem. **78**, 1619–1650 (2006)
16. Revati, K., Pandey, B.: Microbial synthesis of iron-based nanomaterials—a review. Bull. Mater. Sci. **34**, 191–198 (2011)
17. Roh, Y., et al.: Metal reduction and iron biomineralization by a psychrotolerant Fe (III)-reducing bacterium, *Shewanella* sp. strain PV-4. Appl. Environ. Microbiol. **72**, 3236–3244 (2006)
18. Roh, Y., et al.: Extracellular synthesis of magnetite and metal-substituted magnetite nanoparticles. J. Nanosci. Nanotechnol. **6**, 3517–3520 (2006)
19. Sayago, I., et al.: Preparation of tin oxide nanostructures by chemical vapor deposition. In: Tin Oxide Materials, pp. 247–280. Elsevier, Amsterdam
20. Schauermann, S., et al.: Nanoparticles for heterogeneous catalysis: new mechanistic insights. Acc. Chem. Res. **46**, 1673–1681 (2013)
21. Sharma, N., et al.: Preparation and catalytic applications of nanomaterials: a review. RSC Adv. **5**, 53381–53403 (2015)
22. Srivastava, P., et al.: Synthesis of silver nanoparticles using haloarchaeal isolate *Halococcus salifodinae* BK 3. Extremophiles **17**, 821–831 (2013)
23. Wang, Y, Li, Y.: Template-free preparation and photocatalytic and photoluminescent properties of Brookite TiO$_2$ hollow spheres. J. Nanomater. (2019)
24. Wong, S.L.: Chemical vapor deposition growth of 2D semiconductors. In: 2D Semiconductor Materials and Devices, pp. 81–101. Elsevier, Amsterdam
25. Wu, R., et al.: Recent progress in synthesis, properties and potential applications of SiC nanomaterials. Prog. Mater Sci. **72**, 1–60 (2015)
26. Yeary, L.W., et al.: Magnetic properties of biosynthesized magnetite nanoparticles. IEEE Trans. Magn. **41**, 4384–4389 (2005)
27. Zhu, J., et al.: Waste utilization of synthetic carbon quantum dots based on tea and peanut shell. J. Nanomater. (2019)

28. Dowling, A.P.: Development of nanotechnology. Mater. Today 30–35 (2004)
29. Zhu, H., et al.: Uptake, translocation, and accumulation of manufactured iron oxide nanoparticles by pumpkin plants. J. Environ. Monit. 10(6), 713–717 (2008)
30. Tolochko, N.K.: History of nanotechnology. In: Nanoscience and Nanotechnologies. Eolss Publishers, Oxford. https://www.eolss.net
31. Roy, D.N., et al.: Nanomaterial and toxicity: what can proteomics tell us about the nanotoxicology? Xenobiotica 47, 632–643 (2017)
32. Brunner, T.J., et al.: In vitro cytotoxicity of oxide nanoparticles: Comparison to asbestos, silica, and the effect of particle solubility. Environ. Sci. Technol. 40, 4374–4381 (2006)
33. Cassee, F.R., et al.: Particulate matter beyond mass: recent health evidence on the role of fractions, chemical constituents and sources of emission. Inhalation Toxicol. 25, 802–812 (2013)
34. Lahde, A., et al.: In vitro evaluation of pulmonary deposition of airborne volcanic ash. Atmos. Environ. 70, 18–27 (2013)
35. Murr, L.E., Guerrero, P.A.: Carbon nanotubes in wood soot. Atmosph. Sci. Lett. 7:93–95 (2006)
36. Wu, C.Y., et al.: Formation and characteristics of biomimetic mineralo-organic particles in natural surface water. Sci. Rep. 6 (2016)
37. Blanco-Andujar, C., et al.: Elucidating the morphological and structural evolution of iron oxide nanoparticles formed by sodium carbonate in aqueous medium. J. Mater. Chem. 22, 12498–12506 (2012)
38. Cho, M.H., et al.: Redox-responsive manganese dioxide nanoparticles for enhanced MR imaging and radiotherapy of lung cancer. Front. Chem. 5 (2017)
39. Song, S.Q., et al.: Facile synthesis of Fe_3O_4/MWCNTs by spontaneous redox and their catalytic performance. Nanotechnology 21 (2010)
40. Santillo, D., et al.: Microplastics as contaminants in commercially important seafood species. Integr. Environ. Assess. 13, 516–521 (2017)
41. Cole, M., et al.: Microplastics as contaminants in the marine environment: a review. Mar. Pollut. Bull. 62, 2588–2597 (2011)
42. Pawlak, J., et al.: Fate of platinum metals in the environment. J. Trace Elements Med. Biol. 28247–28254
43. Zimmermann, S., Sures, B.: Significance of platinum group metals emitted from automobile exhaust gas converters for the biosphere. Environ. Sci. Pollut. Res. 11, 194–199 (2004)
44. Griffin, S., et al.: Natural nanoparticles: a particular matter inspired by nature: a review. Antioxidants 7(3), 1–21 (2018)
45. Hatami, M., et al.: Engineered nanomaterial-mediated changes in the metabolism of terrestrial plants: a review. Sci. Total Environ. 571, 275–329 (2016)
46. Ferin, J.O. et al.: Increased pulmonary toxicity of ultrafine particles I. Particle clearance, translocation, morphology. J. Aerosol Sci. 21, 381–384 (1990)
47. Oberdörster, G.: Significance of particle parameters in the evaluation of exposure-dose-response relationships of inhaled particles. Inhalation Toxicol. 8, 73–81 (1996)
48. Mossman, B.T., et al.: Asbestos: scientific developments and implications for public policy. Science 247(4940), 294–301 (1990)
49. Zhang, W.: Nanoscale iron particles for environmental remediation: an overview. J. Nanopart. Res. 5, 323–332 (2003)
50. Nanotechnologies: a preliminary risk analysis on the basis of a preliminary workshop in Brussels on 1–2 March by the Health and Consumer Protection Directorate General of the European Commission, European Commission, 2004. www.europa.eu.int/comm/health/ph_risk/documents/ev_20040301_en.pdf. www.oecd.org/chemicalsafety/nanosafety/44108334.pdf. Accessed 05 May 2020

Types and Classification of Nanomaterials

Diana Sannino

Abstract Nanomaterials are currently of utmost importance due to the technological advancements brought from their tailored physical, chemical, and biological properties which lead to superior performance concerning their bulk counterparts. The huge amount of different nanomaterials present in the industrial world and actual research literature requires rationalization with regard to their main characteristics. The several nano-objects belonging to the nanomaterials can be distinguished by their size, kind, shape, aggregate, or agglomerate structure, for instance. The classification of nanoscale materials involving their definition is the subject of this chapter. The main categorizations considered have been (i) the dimensionality of nanomaterials related to their size ranging between 1 and 100 nm, (ii) the individuation of nanomaterials by the type considering the main substances that compose the nanomaterial, (iii) the relevance of morphology for the classification in which full nanomaterials, hollow nanomaterials, nanosheets, nanomaterial of complex architecture, core–shell nanoparticles and nanomembranes can be distinguished, and finally (iv) the classification of nanomaterials by the origin of production.

Keywords Bottom-up approach · Top-down approach · Nanomaterials · Nanotechnology · Nanoscale · Nanodimensions · Nanomorphology · Nanostructures

1 Introduction

'There's plenty of room at the bottom' …this sentence, pronounced by physics Nobel laureate Richard P. Feynman in his lecture at the meeting of the American Physical Society in December 1959 [1], individuated for first the field of nanoscale. Feynman's idea was of manipulating materials at a level of 10^{-9} m, near to the atomic scale, opening to the successive progresses in physics, chemistry, and biology. "Nano" used as a prefix describes "one billionth" of a dimension, so nanotechnology is linked to the

D. Sannino (✉)
Department of Industrial Engineering, University of Salerno, via Giovanni Paolo II 132, 84084 Fisciano, SA, Italy
e-mail: dsannino@unisa.it

nanoscale, where the operations are at the level of few groups of atoms or molecules, with respect to the huge number contained in a mole.

The term "nanotechnology" has been coined in 1974 by Norio Taniguchi to describe the ultrafine dimensions manipulation, introducing also the concept of the "top-down approach" that designate the methods to fabricate miniaturized integrated circuits, optoelectronic devices, mechanical devices, and computer memory devices by a major scale, for example with precision machining with a tolerance of a micron or less. By contrast, the "bottom-up approach" was introduced ten years later by K. Eric Drexler considering the build-up of larger objects from their atomic and molecular components, individuating the forthcoming evolutions of nanotechnology [2]. So the nanotechnologies involve the generation and manipulation of materials at the nanometric scale, either by starting to aggregate groups of atoms (bottom-up approach) or by refining or reducing bulk materials (top-down approach) [3]. Another definition that was given about nanotechnology is the comprehension and control of matter, in size between 1 and 100 nm, to profit from enhanced phenomena for novel applications [4]. A further definition is that nanotechnology involves the development and use of devices and nanomaterials that have a size of only a few nanometers. However, nanotechnology can be additionally individuated as the world of invisible small objects that are controlled by forces of physics and chemistry that cannot be applied at the macro- or human-scale level [5]. Indeed the classic laws of science are different at the nanoscale. One important and prevalent property of the nanoparticles is the large surface areas and essentially no inner mass, which behaves that their surface-to-mass ratio is extremely high (Table 1).

Table 1 Solid particle size and the fraction of atoms located at the particle surface

Number of atoms in a side	Number of atoms at the surface	Total number of atoms	Ratio of surface atoms to the total (%)	Example of particle size and powder
2	8	8	100	
3	26	27	97	
4	56	64	87.5	
5	98	125	78.5	
10	488	1,000	48.8	2 nm
100	58,800	10^6	5.9	20 nm (colloidal silica)
1,000	6×10^6	10^9	0.6	200 nm (titanium dioxide)
10,000	6×10^8	10^{12}	0.06	2 μm (light calcium carbonate)
100,000	6×10^{10}	10^{15}	0.006	20 μm (chalk)

Reproduced from [6]

Fig. 1 Example of the increase of exposed area by reducing the size of a geometrical regular particle. From [3]

In Fig. 1, it is possible to see a cube 1 cm large, with a mass of 1 g that exposes an area of $6 \times 1 cm^2$. Fragmenting this cube in cubettes of 1 mm in size, the total surface area increases to 60 cm². If the size of one cubette is reduced to 10^{-6} mm, i.e., 1 nm, the total surface area reaches the value of 60,000,000 cm², corresponding to 6000 m². This fragmentation could be reached for instance by reducing the size of the material through a machine.

Nanotechnology is a modern and recent multidisciplinary field involving emerging applications in several scientific domains, such as materials, physics, chemistry, biology, engineering, and medicine.

However, the production of nanoscale objects dates to Roman times, in the pre-Christian era. An example is the "Lycurgus cup" [7] held at the London British Museum, where precious metal (gold) of nanometric dimensions was introduced in glass manufacturing. These nanoparticles in the cup display a different color as a function of the illumination, being green if the cup is illuminated externally or red if illuminated internally. However, it must be waited up to 1857 to have the report of the synthesis of colloidal gold and understanding of the nanoparticles interaction with light by Faraday [8].

Nanoparticles were found in the tissues of various organisms, such as bacteria, algae, insects, birds, and mammals. In 1975, Blakemore [9] identified biochemically precipitated magnetite (Fe_3O_4), sensitive to the Earth's magnetic field and used by humans to sense for orientation and navigation.

In the last century, colloid science has been innovated and has been used to obtain fine dispersion of many materials, such as metals, oxides, and organic products [10].

In the early 1960s, Stephen Papell et al. developed magnetic colloidal systems [11] based on finely divided particles of magnetite suspended in paraffin, using oleic acid as a dispersing agent to prevent particle–particle agglomeration or sedimentation. Subsequently, similar magnetic suspensions of different nanometer-sized particles of pure elements (iron, nickel, and cobalt) have also been prepared, in a wide range of carrier liquids [12].

The history of the evolution of nanotechnology has found a vast number of developments in recent years.

The core of nanotechnology is that then when an ordinary material is reduced to the nanoscale demonstrates novel and unpredictable characteristics: strength, chemical reactivity, electrical conductivity, superparamagnetic behavior are examples of the characteristics that are improved than those possessed at the micro- or macroscale.

Nanomaterials are currently being produced at an industrial scale, but strong efforts in research and development of innovative nanomaterials and nanodevices at smaller scales are running.

Drug development, water decontamination, information and communication technologies, and the production of stronger and lighter materials for construction and aeronautics are some areas in which nanotechnology is now retained to bring its interesting advancements.

Nanomaterials are size dependent. As pointed out below, the properties of the materials change as their size approaches the nanoscale and as the percentage of the atoms at the surface of the material becomes significant.

For the sake of comparison, bulk materials, mainly particles larger than one micrometer contain an insignificant percentage of atoms at the surface concerning the number of atoms in the bulk of the same material. And, therefore, they do not behave or exhibit size-dependent changes in their physical and chemical properties.

Nanomaterials can possess unforeseen optical properties since they are small enough to confine their electrons and produce quantum effects. Within this size range, nanoparticles can bridge the gap between small molecules and bulky materials in terms of energy states [13].

The studied and applied nanomaterials are now a huge variety, and a lot of researches are running actually. This vast assortment needs logical classifications, and distinctions are considered according to the main properties of the nanomaterials.

2 Classification of Nanomaterials

There are several ways of nanoparticle classifications.

The first classification arises from the source of nanomaterials, which could be natural or man-made. Another difference comes from the fact if they are the desired final products or by-products, furthermore, they can be generally organic or inorganic. Regarding the first point, nanoparticles are a product of modern technology, but they are also generated in natural events such as volcano eruptions, forest fires, dried salted aerosols, or ultrafine sand grains of mineral origin (e.g., oxides, carbonates). Some ultrafine particles are unintentionally created by the combustion of fossil fuels or during barbecuing and are products of man activities.

A simple categorization depends on the type of nanomaterials [14], i.e., (i) carbon-based nanomaterials, (ii) inorganic-based nanomaterials, (iii) organic-based nanomaterials excluding the carbon nanomaterials, and (iv) composite-based nanomaterials.

Furthermore, nanomaterials can be classified with respect to their dimensionality (size), morphology (shape), surface properties composition, uniformity, functionalization, and agglomeration [15].

For the latter point, nanotechnology products can be distinguished on the base of their appearance: they could be powders, dispersions (powders in liquids), coatings (thin film covering surface), macrosolids (e.g., interconnecting solid particles through nanoparticles, famous as nanowhiskers).

In the framework of nanotechnology, the term "nano" refers almost exclusively to particle length. According to the definition of the EU-Commission SCENHIR2, the nanoworld is restricted to those "objects whose extension in all three dimensions lies between 1 and 100 nm [16]. Those that extend in only two dimensions on the nm-scale are termed nanotubes and particulate objects; those with only a single dimension under 100 nm are termed nanopellets."

In the following, some properties will be better evidenced for the classification, starting on the most characteristic, that is dimensionality.

2.1 Dimensionality of Nanomaterials

A nanostructure can be described and classified on the basis of the number of dimensions that belong to the nanoscale, according to the definition given in 2007 by Pokropivny and Skorokhod [17]. Of course, to be a nanomaterial, an object has to have at least one dimension in the range of 1–100 nm.

In describing nanostructures, it is needed to differentiate between the number of dimensions that material possesses on the nanoscale.

Nanoclusters are atoms or molecules groups, 1–100 nm in each spatial dimension. These small groupings of atoms and molecules contain few to some thousands of units and have size mostly in the single nanometer scale (Fig. 2).

Nanoclusters are usually distinguished from nanoparticles. Instead, the term nanoparticle is often used when speaking of bigger clusters with diameters from several nanometers to several hundreds of nanometers.

Quantum dots (QDs) are synthetic nanoscale crystals that can transfer electrons and show a variety of properties, depending on their composition and shape. They are semiconductors and emit particular colors when irradiated by UV light (Fig. 3).

These nanostructures are individuated in the literature **as zero dimensional nanostructures (0D)**.

They show unique physical properties, allowing much higher absorption of solar radiation. Other size-dependent properties change involves quantum confinement in semiconductor particles, surface plasma resonance in some metal nanoparticles, and chemical reactivity that can be utilized for image formation in the photography field.

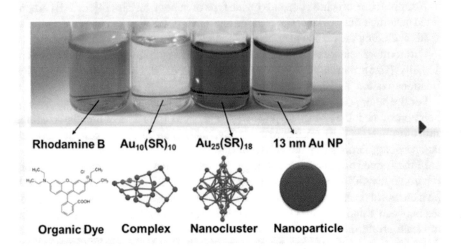

Fig. 2 Differences between organic dye molecules, a gold complex, a nanocluster of gold, and a nanoparticle (Rhodamine B, Au10 complex, Au25 nanocluster of 1 nm metalcore, 13 diameter Au nanoparticles. Courtesy of [18]

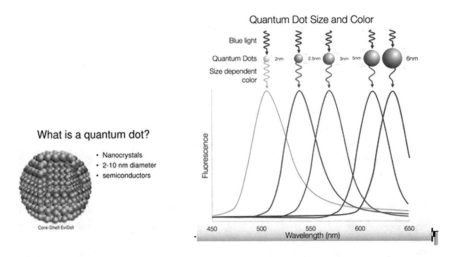

Fig. 3 Quantum dot scheme and interaction with blue light with the emission of different colors in dependence of size. From [19]

The nanostructures that have two characteristic dimensions between 1 and 100 nm and one no-nanometer size are categorized **as one-dimensional nanostructures (1D).**

Nanotubes, nanofibers, and nanowires are examples of these one-dimensional nanostructures. Indeed, the length of a nanotube and nanowires could be much greater

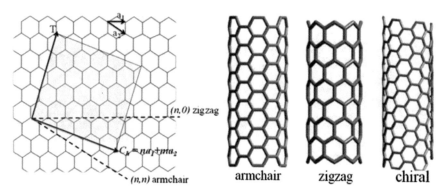

Fig. 4 Armchair, zigzag and chiral types of carbon nanotubes obtained by rolling a graphene sheet in different directions. Source [20]

than 100 nm. The nanotubes can be different according to their chirality (the property of asymmetry dealing with the mirror image of an object if is not superposed onto it) that attributes different, e.g., conduction properties (Fig. 4).

1D nanomaterials can be amorphous or crystalline, be a single crystallite or polycrystalline materials, be of simple or complex composition, be standalone materials or embedded in within another medium, be of course metallic, ceramic, or polymeric.

When there are nanomaterials composed of thin layers that may have a thickness of at least one atomic layer, having many atoms on their surface, they are categorized **as two-dimensional complex materials (2D)**.

2D nanomaterials should be the thinnest nanomaterials due to their low thickness and higher dimensions on macroscale/nanoscale. These nanomaterials possess generally a layered structure with strong in-plane bonds and weak interaction forces (van der Waals) within the layers and can be produced by laminating precursors. The matter of the fact is that, since two of the dimensions are not confined to the nanoscale, 2D nanomaterials show plate-like shapes, such as nanofilms, nanolayers, and nanocoatings, and can be amorphous or crystalline, of various chemical compositions, applied as a single layer or as multilayer structures, used placed on a substrate, be integrated into a complex matrix of materials, and of course be metallic, ceramic, or polymeric.

Graphene, hexagonal boron nitride (hBN), and metal dichalcogenides (MX2) are examples of this category (Fig. 5) [21, 22].

There are also bulk materials with all dimensions above 100 nm and are named **three-dimensional nanostructures (3D).** Materials having a nanocrystalline structure or involving the presence of features at the nanoscale are examples of this dimensionality.

Bulk nanomaterials can be formed by multiple organizations of nanosized crystals, usually found in different orientations. 3D nanomaterials can be composed of dispersions of nanoparticles, bundles of nanowires, and nanotubes as well as multinanolayers, but more complex arrangements are possible. Nanoflowers are for

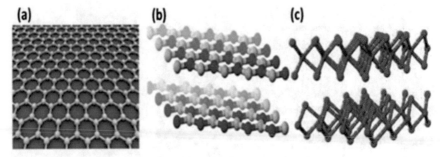

Fig. 5 Structure of **a** single layer of graphene with a lattice of carbon atoms, **b** boron nitride nanosheets with B in blue and N in pink, and **c** tungsten diselenide (WSe2) with W in blue and Se in yellow [21]

instance a kind of possible architectures. A general distinction in the dimensionality of carbon-based nanomaterials is reported in Fig. 6

Some properties of nanomaterials at the nanoscale can be remembered, such as quantum effects: 0D nanomaterials, with all three dimensions below 100 nm, have electrons in 3D space. No electron delocalization, meaning freedom to move, happens. In the case of 1D nanomaterials, electron confinement takes place in 2D and delocalization occurs along the main axis of the elongated nanostructure. With

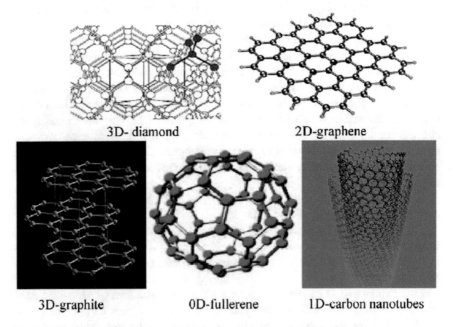

Fig. 6 Examples of different dimensionalities for carbon nanomaterials [23]

regard to 2D nanomaterials, the conduction electrons will be limited through the thickness but delocalized in the plane of the sheets.

2.2 Classification of Nanomaterials by the Type

Most of the current nanomaterials can be grouped into four material-based categories, and they are:

(i) Carbon-based nanomaterials: obviously, these nanomaterials containing carbon can be found in several morphologies and phases. They include fullerenes with a defined number of carbon atoms in the structures (for instance C60) that are ellipsoids or spheres, carbon nanotubes (CNTs) that are wire hollow tubes, carbon nanofibers that have the form of nanowires, the carbon black, usually in the form of particulate, and graphene (Gr) composed by a layer of exfoliated graphite. It is possible also to find carbon onions. The allotropes of carbon are reported in Fig. 7.

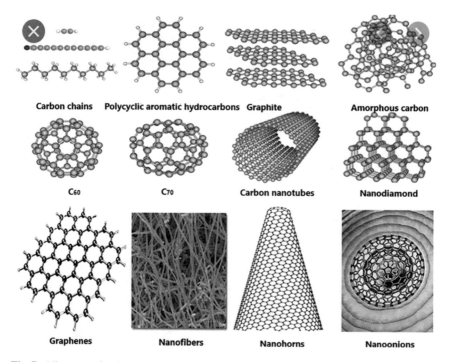

Fig. 7 Allotropes of carbon. Courtesy of [24], cited by [25]

Several methods of production are possible, such as laser ablation, arc discharge, and chemical vapor deposition (CVD) among the most important production technologies [8].

(ii)   Inorganic-based nanomaterials: these nanomaterials are mainly made of metal or metal oxide particles less than 100 nm. They are composed of pure precious metals such as Au or Ag nanoparticles or being in the form of metallic oxides such as TiO_2 and ZnO. This category also includes semiconductors such as silicon and ceramics.

(iii)   Organic-based nanomaterials: excluding carbon-based or inorganic-based nanomaterials, it is possible to individuate nanomaterials and nanostructures composed of organic matter. Dendrimers, micelles, liposomes, and polymer can be obtained by the utilization of non-covalent (weak) interactions, profiting of the self-assembly properties, or design of molecules.

(iv)   Composite-based nanomaterials: they are multiphase complex nanostructures with one phase at least on the nanoscale dimension. They can both combine nanoparticles of different compositions and shape together or include nanoparticles within bulk-type materials (e.g., hybrid nanofibers) or more complicated structures (e.g., metal–organic frameworks). The nanocomposites can be any mixture of carbon-based, metal-based, or organic-based nanomaterials and bulk materials of every kind and form (metal, ceramic, or polymer, etc.).

(v)   Bionanomaterials: DNA, viruses are an example of biological nanomaterials encountered. For example, DNA is a molecular nanowire composed of repeating molecular units, organic in nature.

Nanobiotechnology is a special branch of nanoscience devoted to specifical biological applications. Bionanomaterials include apart from DNA, amyloid fibrils, actin filaments, aromatics peptides, bacteriophages, minerals, viruses, enzymes, and nucleic acids.

2.3   Classification by Morphology

Apart from the previous classification, the several nano-objects belonging to the nanomaterials can be classified by their shape and spatial organization (aggregate or agglomerate structure) since their low dimensions favor the occurring of interactions. Indeed, small particles are very active due to their high surface area and tend to agglomerate (formation of a weak bond among the nanoparticles) or aggregate (formation of stable bonds between the nanoparticles) with each other.

In dependence on the material composition, crystal structure, and manufacturing method, the morphology of nanostructures changes significantly.

The morphological characteristics to be taken into account are the fullness, the flatness, the aspect ratio, and the spatial position of each unit in the case of hybrid nanoparticles. The aspect ratio indicates the relationship between the two main dimensions of a two-dimensional or tri-dimensional figure. A general distinction exists between "high aspect ratio "and "low aspect ratio" particles. The high aspect ratio comprises nanotubes and nanowires, while the small aspect ratio nanoparticles have spherical, oval, cubic, prism, helical, and pillar shapes [13].

Nanoparticles can be amorphous or crystalline, be single crystalline or polycrystalline, be composed of single or multi-chemical elements, be metallic, ceramic, or polymeric and exhibit various shapes and forms. They can exist individually or incorporated into a matrix.

The production of nanoparticles of the same material with a variety of shapes, such as spheres, rods, tubes, needles, cubes, and octahedrons, is allowed by changing synthesis methods and operating conditions. It is interesting example of hydrothermal synthesis, from which (by varying temperature, pressure, reagents concentration, treatment time, and pH) different morphologies, compositions, and crystallinity nanomaterials can be obtained [26].

An example related to the synthesis of hematite is reported in [27]. Nanorhombohedra, nanorods, nanocubes, nanoneedles, and nanobars can be created by varying the synthesis conditions.

The morphological variety of nanomaterials also built with organic molecules is almost countless and can give more complex architectures. For instance, self-assembling methods using DNA as building blocks allow the production of three-dimensional structures sized between 10 and 100 nm, such as polygon frameworks, gears, bridges, and bottles. [28].

Morphological multiplicity permits these materials to contain a large number of surface atoms that individuated their physical and chemical properties. By contrast, the thermodynamic stability of most nanomaterials is to be carefully considered, and their nonequilibrium morphologies, in the distance to the shape of monocrystals for a given substance, correspond to minimums of free energy of the system.

Nanospheres or nanoregular particles are found often in nature and the research products and are interesting to be obtained.

For example, gold can be obtained in the forms of nanocrystals of a spherical or different shape. Nanoparticles pentagonal, triangular, hexagonal, and others can be observed in Fig. 8.

The differences given by each of the diverse-sized nanogold for instance regard the absorption and reflection of the light, which is altered in dependence by size and bonding arrangement of the nanoparticles.

Always following the example of gold nanoparticles, they may easily be incorporated in liquid mediums or found in solid forms [30].

Full nanomaterials of high aspect ratio are nanowires and nanofibers. Figure 9 reports the different forms in which TiO_2 can be obtained by controlling hydrothermal synthesis, nanoparticles, nanofibers nanowires, and nanotubes.

Considering the latter shape, it must be considered that many nanomaterials have a hollow cavity, which is often presented in geometrical regular shapes or elongated shapes. Examples of these nanomaterials are hollow nanospheres, fullerenes, nanoballoons, nanotubes, nanocapsules.

Monodisperse porous hollow nanospheres with superparamagnetism property can be obtained via a hydrothermal reaction based on solid Fe_3O_4 nanospheres (Fig. 10).

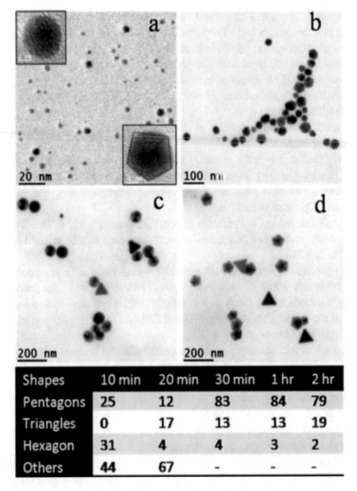

Shapes	10 min	20 min	30 min	1 hr	2 hr
Pentagons	25	12	83	84	79
Triangles	0	17	13	13	19
Hexagon	31	4	4	3	2
Others	44	67	-	-	-

Fig. 8 Temporal evolution of morphology by seeding method **a** seed, **b** 10 min, **c** 20 min, and **d** 30 min grown samples. Inset in (**a**) shows seed morphology at higher magnification. Pentagonal, triangular, hexagonal, and spherical shapes are obtained. Population statistic of shapes is given in the table. Courtesy of [29]

The most famous hollow nanomaterials are the nanotubes. In particular, carbon nanotubes are built by the enrollment of graphite or graphene sheets: In the first case, multiwalled nanotubes are obtained and in the second single-walled nanotubes [33].

The properties and performance of nanotubes are so affected by the diameter, the number of walls, length, chirality, functionalization, Van der Waals forces, and quality [34]. The carbon nanotubes are different from the carbon fibers, whose structure is showed in Fig. 11.

Figure 11 also shows different ways to assemble nanofibers by different stackings of graphite planes.

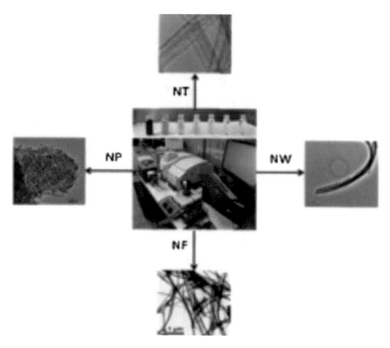

Fig. 9 Synthesis, characterization, and photocatalytic activity of TiO2 nanostructures: Nanotubes (NT), nanofibers (NF), nanowires(NW), and nanoparticles (NP). Courtesy of [31]

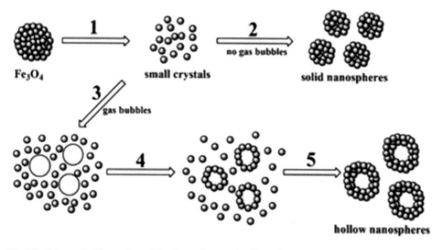

Fig. 10 Schematic illustration of the formation mechanism of porous Fe$_3$O$_4$ hollow nanospheres formed from nanocrystals. Courtesy of [32]

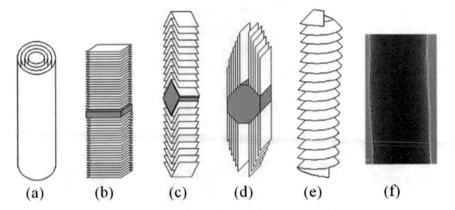

Fig. 11 Structure of **a** multiwalled carbon nanotubes, **b** graphene platelets, **c** graphene fishbones, **d** graphene ribbons, **e** stacked cup carbon nanofibers, and **f** amorphous carbon nanofibers without graphene layers. Courtesy of [35]

Nanoribbon and nanobelt of carbon can be also obtained, as shown in [36].

Nanowhiskers instead are types of filamentary crystals (whisker) with the cross-sectional diameter ranging in the nanoscale and length to diameter ratio >100.

Short lengths can be attributed to the nanomaterials termed nanorods, nano-objects with two dimensions ranging from 1 to 100 nm and the length being slightly greater (see Fig. 12c).

Nanosheets are few-atoms thick layers, with the two other dimensions greater than 100 nm. One example is graphene, composed of one-atom-thick layer of polycyclic carbon.

Indeed, starting from single nano-shapes, complex architecture due to their assembly in regular forms gives rise to nanostructures, as shown in Fig. 12e, f.

Nanomaterials of complex architecture are dendrimers, highly branched, multivalent, and monodisperse polymers, which have received continuous interest in recent years because of their potential application in advanced technology and medicine. The synthesis of dendrimer proceeds according to a divergent method starting from a multifunctional core molecule, to which branched building blocks called monomers are attached up via the sequential steps (Fig. 13) in a concentric way [37]. Dendrimer branches have a common core and are individuated by generation: a first-generation dendrimer possesses one branch point in each branch, and a second-generation dendrimer has two branch points, etc. Dendrimers of the third and further generations have high-density molecular structures and an almost spherical shape.

Carbon nano-onions are a variant of fullerenes, in which multiple shells of spherical carbon can be found. The nanomaterial consists of spherical closed carbon shells, multilayered quasi-spherical, and polyhedral-shaped shells, with a structure resembling that of an onion [38]. They are individuated also like nanoscale carbon structures formed by nested carbon spheres (Fig. 14).

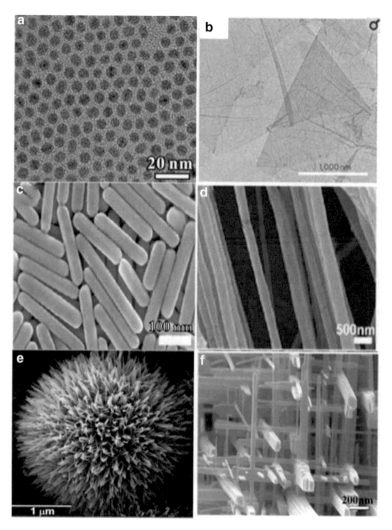

Fig. 12 Nanomaterials with different morphologies: **a**- nonporous Pd NPs (0D), **b**- graphene nanosheets (2D), **c**- Ag nanorods (1D), **d**- polyethylene oxide nanofibers (1D), **e**- urchin-like ZnO nanowires (3D), **f**- WO_3 nanowire network (3D) [14]. Courtesy of [64]

Similar to nano-onions, core–shell nanoparticles are nanocomposite materials with a central core of a substance and an external shell different in composition, and they are usually spherical or pseudospherical but can assume other shapes (Fig. 15). Core–shell nanoparticles differ from previous nanoparticles, composed of a single material, and are biphasic materials concerning the inner core and the outer shell. There could be also the presence of multiple phases assembled as different shells. These particles are of interest because they show unique characteristics arising from the combination of core and shell material, geometry, and design. For example,

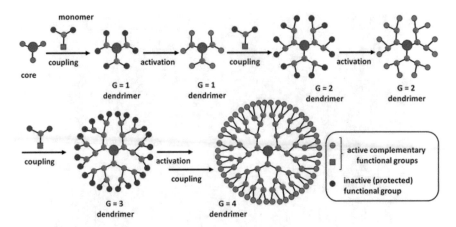

Fig. 13 Synthesis of dendrimers according to the divergent method. Courtesy of [37]

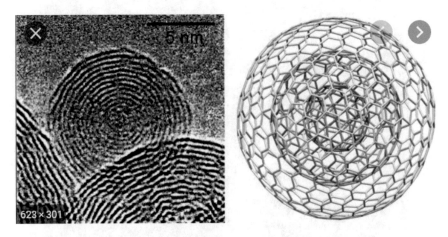

Fig. 14 Carbon nano-onion TEM image and structure. Courtesy of [39]

nanospheres and microspheres containing a magnetic core embedded in a non-magnetic matrix are used in numerous environmental and biological applications [40]. They can be (i) organic, for instance with an external polymeric shell and an inner metallic core, or (ii) inorganic, further categorized into two main classes, namely (a) silica-based core–shell nanoparticles and (b) metal-based core–shell nanoparticles.

Biphasic materials are also nanocapsule. They differ from core–shell particles because they bring drug active principia in their interior. Usually, the other phase present is a polymer. Drug-charged nanospheres are dissimilar by a nanocapsule because the latter enclose the drug in the core.

It is interesting to note that also void space may have nanodimensions. Nanopores can be defined in each material and found together with the more consolidated

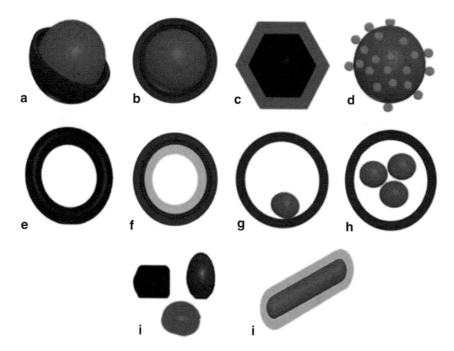

Fig. 15 Schematic (A-J) pictures of the different structure of core–shell nanoparticles: **a** core–shell nanoparticles; **b** core double-shell particles or core multi-shell nanoparticle; **c** polyhedral core–shell nanoparticles; (D) core porous-shell nanoparticles; **e** hollow core–shell nanoparticles or single-shell nanoparticles; **f** hollow-core double-shell nanoparticles; **g** moveable core–shell nanoparticles; **h** multi-core–shell nanoparticles; **i** irregular shape core–shell nanoparticles; and **j** rod core–shell nanoparticle. Courtesy of [41]

definition of micro- (pores diameter less than 2 nm) meso- (2–50 nm) and macro- (pores diameter more than 50 nm) porosity.

Considering that the aggregation or agglomeration of the nano-objects, in general, creates pores of similar dimensions, nanopores can be easily individuated.

Among the nanoporous materials, it has to mention the nanomembranes.

Nanomembranes are defined as filters that separate liquids and gases at the molecular level. They are constituted by freestanding structures with a thickness in the range of 1–100 nm and an extremely large aspect ratio, of at least a few orders of magnitude and bring some nanoporosity. Nanomembranes can be composed of single nanomaterials or associated with other substance in composite membranes (Fig. 16).

Interesting is the macro-form of nanomaterials. A compact and lightweight compressed nanotubes ensemble gives rise to the so-called buckypaper handle sheets. Nanotubes usually adhere together to form bundles or strands. Nanotube super fiber materials are the fibrous assembly of nanotubes and strands [34].

The list of possible nanomorphology is very long; however, we focused on the most found forms of nanomaterials.

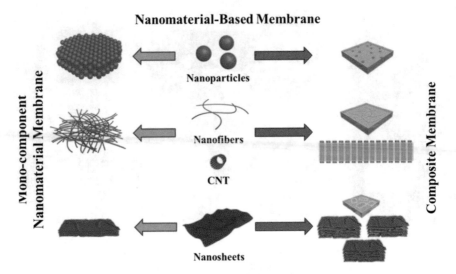

Fig. 16 Nanomaterials organization to form nanomembranes as aggregates of nanoparticles, nanofibers, and nanosheets [42]

2.4 Classification of Nanomaterials by the Origin

In this classification of nanomaterials, three main categories based on their origin can be found:

(i)   incidental nanomaterials, which are products found incidentally or as by-products of chemical processes. For instance, to this category belongs the nanoparticles produced from vehicle engine exhaust and anthropic or natural combustion processes such as forest fires;

(ii) engineered nanomaterials, produced by human manufacturing with defined properties for desired final applications.

(iii) naturally produced nanomaterials, which can be synthesized in the bodies of biorganisms, flora, fauna, and human bodies.

However, the distinctions between naturally occurring, incidental, and manufactured nanoparticles are often indistinct. In some cases, incidental nanomaterials can be reflected as a subgroup of naturally produced nanoparticles.

2.4.1 Incidental Nanomaterials

Incidental nanomaterials are present in all natural environments such as ground and surface waters, marine waters, soils, or the atmosphere. The main difference between the incidental and the engineered nanomaterials is that the morphology is not controlled at all. Nanoparticles in nature are produced from photochemical reactions, dust storms, volcanic eruptions, forest fires, but they are also fragments of skin and hair from the plants and animals.

The nanoparticles produced by volcanic eruptions and forest fires are of natural origin, but they are produced in huge amounts, significantly affecting the air quality. In the same way, combustion processes for transportation, industrial operations, and charcoal burning are some of the human sources that induce the presence of hazardous nanoparticles in the environment.

As nanoparticles, it is possible also to consider atmospheric aerosols, 90% of which are generated by natural phenomena and about 10% by human activity [43]. Dust storms in the desert and terrestrial regions are the main source of natural nanoparticles, and these nanoparticles can migrate from a terrestrial region to another, transferring nano- and micro-sized minerals to thousands of kilometers. About 50% of the atmospheric particles originate from dust storms in deserts and they are sized in the range of 100–200 nm [44].

Also, cosmic dust is a kind of incidental nanomaterial, composed of a large assortment of carbide, oxide, nitride, silicate, carbon, nanodiamonds, and also organic-based nanoparticles [45]. The most frequent processes that induce the formation of nanomaterials in space are electromagnetic radiation, pressure gradients, dramatic temperature, physical collisions, and shock waves.

Automobile exhausts are a source of particulates, 20–130 nm-sized from diesel engines and 20–60 nm-sized particles from gasoline [46]. Also, CNTs and fibers are released as by-products during diesel and gas combustion processes [47]. The results suggested that humans are routinely exposed to carbon nanotubes.

Among the sources of incidental nanomaterials can be cited cigarette smoking and building demolition that lead to the spread of nanoparticles into the atmosphere [48].

Sea salt aerosols are a different type of natural nanoparticles obtained from the froth of the seas and oceans waves into the atmosphere by evaporation-mediated natural precipitation [49], being typically in size from 100 nm to few micrometers.

2.4.2 Engineered Nanomaterials

Engineered nanoparticles are intentionally formed and designed with very specific properties starting from their shape, size, surface properties, and chemistry. They are often aerosols, colloids, or powders. The properties of nanomaterials may depend more on the surface area than the chemistry of the nanoparticles itself. Their intentional production started in the half past century, leading to obtained amounts relative to the other two types of nanomaterials. New applications are continuing to develop.

The most common engineered nanomaterials are nanoscale metal oxides, (of which the most diffused are titanium dioxide, iron, and aluminum oxides) nanoscale polymers, and polymeric nanocomposite materials that are used in the buildings construction, electronics, computer, pharmaceutical, and medical industries. They are manufactured from kilograms to tons.

Quantum dots, nanoshells, nanocages, and nanobranches are now considered as emerging engineered nanomaterials being currently used in new photovoltaic cells, as drug delivery nanovehicles or immunological sensing devices [50]. Because of the variety of engineered nanomaterials, it is difficult to predict the long-term trends of their industry.

Engineered nanomaterials are present in many commercial products of common use. For example, they are included in some electrical appliances, medicines, cleaning products, cosmetics, and personal care products (such components of sunscreens and toothpaste), paints and building materials, textiles (i.e.,mainly to get an antibacterial effect), sporting goods, pollution control applications.

It must be considered also that chemical manufacturing, welding, ore refining, and smelting produce nanoparticles. Biomedical and healthcare products or commercial cosmetics can contain nanoparticles intentionally produced such as carbon nanoparticles [51], TiO_2 nanoparticles [52], and hydroxyapatites [53]. For example, titanium oxide nanoparticles bigger in size than 100 nm are broadly exploited as active pigments in cosmetic creams and sunscreens [54]. Air sanitizer sprays, wet wipes, food storage containers, shampoos, and toothpaste can, similarly, contain Ag nanoparticles [55]. Although there is an emerging growth of products that included nanomaterials, their hazardous effects on human health are largely unknown.

2.4.3 Naturally Produced Nanomaterials

Naturally produced nanomaterials nanoparticles and nanostructures are present in all living organisms including viruses. Nanoparticles are accumulated from bacteria, algae, and viruses and can be found also in more complex organisms: plants, insects, birds, animals, and humans include nanoparticles. Their identification helps to understand the mechanism of formation: for instance, plants utilize microelements available in soil and water and accumulate them in the forms of nanoparticles. About the nano-organisms, plant viruses, nanobes, and actinomycetes can bind to soluble metals and to precipitate them forming nanoparticles of several metals, for instance of Ag [56], gold [57], alloys [58], non-magnetic oxides [59], magnetic oxide [60], metal sulfide quantum dots [61].

Also, fungi contain a variety of enzymes and are simple to handle, which gives the possibility of preparation of various sized and shaped nanoparticles.

It is worthwhile to mention that in the plants, nanoparticles and nanostructures are present, such as natural fibers composed of cellulosic-fibrils at the nanoscale level, and moreover, plants surfaces, especially leaves, contain nanostructures that are used for numerous purposes; the most famous is the so-called lotus effect that gives self-cleaning and super-wettability of the leaves [62].

Nanoparticles and nanostructures are also inserted in insects. Insect wing surfaces expose an irregular or highly ordered micro- and nano-structure to shield them against wetting and pollutants. Besides, the literature reveals that they possess multilayered nanostructures acting as diffraction gratings, which produce interference and therefore iridescence [63].

Nanoparticles and nanostructures in animals and birds are of a large variety. It is interesting to point out that the human body contains nanostructures fundamental for the metabolism of the body, such as bones, enzymes, proteins, antibodies, and DNA [14].

3 Conclusions and Future Perspective

The kinds of nanomaterials are wide and different in types, sizes, nanostructures, and sources. They can occur naturally or be chemically, mechanically, physically, or biologically synthesized, with various compositions and nanostructures. The most important distinction is on the base of their size; however, further classifications help to understand how to build nanomaterials and to catch a glimpse of their potential applications or toxicity effects. It should be said that some nanomaterials might fall on the borders of any of these categories.

3D hierarchical structures can be derived from the organization of nanomaterials in complex structures and are attractive for several applications and the manipulation of nanomaterials.

An additional distinction could be based on the functionalization of the nanomaterials; however, the large variety of possibility yields at the moment difficult their categorization.

References

1. Feynman, R.P.: There's plenty of room at the bottom. California Institute of Technology, Engineering and Science Magazine (1960)
2. Corbett, J., McKeown, P., Peggs, G., Whatmore, R.: Nanotechnology: international developments and emerging products. CIRP Ann. **49**(2), 523–545 (2000)
3. www.nano.gov. NNI
4. https://www.sciencedirect.com/topics/engineering/taniguchi
5. Theodore, L., Kunz, R.G.: Nanotechnology: Environmental Implications and Solutions. Wiley Online Library, New York (2005)
6. Yokoyama, T.: Nanoparticle Technology Handbook. Elsevier, Amsterdam (2012)
7. www.britishmuseum.org/explore/highlights/highlight_objects/pe_mla/t/the_lycurgus_cup. aspx
8. Faraday, M.: X. The Bakerian Lecture.—Experimental relations of gold (and other metals) to light. Philos. Trans. R. Soc. Lond. **147**, 145–181 (1857)
9. Blakemore, R.: Magnetotactic bacteria. Science **190**(4212), 377–379 (1975)
10. Luttge, R.: Microfabrication for Industrial Applications. William Andrew (2011)
11. Stephen, P.S.: Low viscosity magnetic fluid obtained by the colloidal suspension of magnetic particles. Google Patents (1965)
12. Kaiser, R., Miskolczy, G.: Magnetic properties of stable dispersions of subdomain magnetite particles. J. Appl. Phys. **41**(3), 1064–1072 (1970)
13. Benelmekki, M.: An introduction to nanoparticles and nanotechnology. In: Designing Hybrid Nanoparticles. Morgan & Claypool Publishers (2015)

14. Jeevanandam, J., Barhoum, A., Chan, Y.S., Dufresne, A., Danquah, M.K.: Review on nanoparticles and nanostructured materials: history, sources, toxicity and regulations. Beilstein J. Nanotechnol. **9**(1), 1050–1074 (2018)
15. Bowman, D.M.: More than a decade on: mapping today's regulatory and policy landscapes following the publication of nanoscience and nanotechnologies: opportunities and uncertainties. NanoEthics **11**(2), 169–186 (2017)
16. Leitgeb, N., Auvinen, A., Danker-hopfe, H., Mild, K.: SCENIHR (Scientific Committee on Emerging and Newly Identified Health Risks). Potential health effects of exposure to electromagnetic fields (EMF). Scientific Committee on Emerging and Newly Identified Health Risks SCENIHR Opinion on Potential health 10:75635 (2016)
17. Pokropivny, V., Skorokhod, V.: Classification of nanostructures by dimensionality and concept of surface forms engineering in nanomaterial science. Mater. Sci. Eng., C **27**(5–8), 990–993 (2007)
18. Zhou, M., Zeng, C., Li, Q., Higaki, T., Jin, R.: Gold nanoclusters: bridging gold complexes and plasmonic nanoparticles in photophysical properties. Nanomaterials **9**(7), 933 (2019)
19. https://physicsopenlab.org/2015/11/18/quantumdots/.
20. https://www.intechopen.com/books/carbon-nanotubes-polymernanocomposites/polymer-car bon-nanotube-nanocomposites.
21. Rafiei-Sarmazdeh, Z., Zahedi-Dizaji, S.M., Kang, A.K.: Two-dimensional nanomaterials. In: Nanostructures. IntechOpen (2019)
22. Zhang H (2015) Ultrathin two-dimensional nanomaterials. ACS Nano **9**(10), 9451–9469
23. Chen, Y., Zhang, B., Liu, G., Zhuang, X., Kang, E.-T.: Graphene and its derivatives: switching ON and OFF. Chem. Soc. Rev. **41**(13), 4688–4707 (2012)
24. Tripathi, A.C., Saraf, S.A., Saraf, S.K.: Carbon nanotropes: a contemporary paradigm in drug delivery. Materials **8**(6), 3068–3100 (2015)
25. Kharisov, B.I., Kharissova, O.V.: Carbon Alotropes: Metal-Complex Chemistry, Properties and Applications. Springer, Berlin (2019)
26. https://eng.thesaurus.rusnano.com/wiki/article1233.
27. Pu, Z., Cao, M., Yang, J., Huang, K., Hu, C.: Controlled synthesis and growth mechanism of hematite nanorhombohedra, nanorods and nanocubes. Nanotechnology **17**(3), 799 (2006)
28. Dietz, H., Douglas, S.M., Shih, W.M.: Folding DNA into twisted and curved nanoscale shapes. Science **325**(5941), 725–730 (2009)
29. Singh, M.K., Mukherjee, B., Mandal, R.K.: Growth morphology and special diffraction characteristics of multifaceted gold nanoparticles. Micron **94**, 46–52 (2017)
30. Devi, G.K., Suruthi, P., Veerakumar, R., Vinoth, S., Subbaiya, R., Chozhavendhan, S.: A Review on metallic gold and silver nanoparticles. Res. J. Pharmacy Technol. **12**(2), 935–943 (2019)
31. Camposeco, R., Castillo, S., Navarrete, J., Gomez, R.: Synthesis, characterization and photocatalytic activity of TiO2 nanostructures: nanotubes, nanofibers, nanowires and nanoparticles. Catal. Today **266**, 90–101 (2016)
32. Liu, X., Li, Y., Zhu, W., Fu, P.: Building on size-controllable hollow nanospheres with superparamagnetism derived from solid Fe_3O_4 nanospheres: preparation, characterization and application for lipase immobilization. CrystEngComm **15**(24), 4937–4947 (2013)
33. Hirsch, A., Vostrowsky, O.: Functionalization of carbon nanotubes. In: Functional Molecular Nanostructures, pp. 193–237. Springer, Berlin (2005)
34. Schulz, M.J., Ruff, B., Johnson, A., Vemaganti, K., Li, W., Sundaram, M.M., Hou, G., Krishnaswamy, A., Li, G., Fialkova, S.: New applications and techniques for nanotube superfiber development. In: Nanotube Superfiber Materials, pp. 33–59. Elsevier, Amsterdam (2014)
35. Zhang, B., Kang, F., Tarascon, J.-M., Kim, J.-K.: Recent advances in electrospun carbon nanofibers and their application in electrochemical energy storage. Prog. Mater Sci. **76**, 319–380 (2016). https://doi.org/10.1016/j.pmatsci.2015.08.002
36. Sau, T.K., Rogach, A.L., Jäckel, F., Klar, T.A., Feldmann, J.: Properties and applications of colloidal nonspherical noble metal nanoparticles. Adv. Mater. **22**(16), 1805–1825 (2010)
37. Sowinska, M., Urbanczyk-Lipkowska, Z.: Advances in the chemistry of dendrimers. New J. Chem. **38**(6), 2168–2203 (2014)

38. Bartelmess, J., Giordani, S.: Carbon nano-onions (multi-layer fullerenes): chemistry and applications. Beilstein J. Nanotechnol. **5**(1), 1980–1998 (2014)
39. Camisasca, A., Giordani, S.: Carbon nano-onions in biomedical applications: promising theranostic agents. Inorg. Chim. Acta **468**, 67–76 (2017)
40. Krishnan, K.M.: Biomedical nanomagnetics: a spin through possibilities in imaging, diagnostics, and therapy. IEEE Trans. Magn. **46**(7), 2523–2558 (2010)
41. Khatami, M., Alijani, H.Q., Nejad, M.S., Varma, R.S.: Core@ shell nanoparticles: greener synthesis using natural plant products. Appl. Sci. **8**(3), 411 (2018)
42. Ying, Y., Ying, W., Li, Q., Meng, D., Ren, G., Yan, R., Peng, X.: Recent advances of nanomaterial-based membrane for water purification. Appl. Mater. Today **7**, 144–158 (2017)
43. Taylor, D.A.: Dust in the wind. Environ. Health Perspect. **110**(2), A80–A87 (2002)
44. d'Almeida, G.A., Schütz, L.: Number, mass and volume distributions of mineral aerosol and soils of the Sahara. J. Climate Appl. Meteorol. **22**(2), 233–243 (1983)
45. Barnard, A.S., Guo, H.: Nature's Nanostructures. CRC Press, Boca Raton (2012)
46. Westerdahl, D., Fruin, S., Sax, T., Fine, P.M., Sioutas, C.: Mobile platform measurements of ultrafine particles and associated pollutant concentrations on freeways and residential streets in Los Angeles. Atmos. Environ. **39**(20), 3597–3610 (2005)
47. Soto, K., Carrasco, A., Powell, T., Garza, K., Murr, L.: Comparative in vitro cytotoxicity assessment of some manufacturednanoparticulate materials characterized by transmissionelectron microscopy. J. Nanopart. Res. **7**(2–3), 145–169 (2005)
48. Stefani, D., Wardman, D., Lambert, T.: The implosion of the Calgary General Hospital: ambient air quality issues. J. Air Waste Manag. Assoc. **55**(1), 52–59 (2005)
49. Buseck, P.R., Pósfai, M.: Airborne minerals and related aerosol particles: effects on climate and the environment. Proc. Natl. Acad. Sci. **96**(7), 3372–3379 (1999)
50. Kahru, A., Dubourguier, H.-C.: From ecotoxicology to nanoecotoxicology. Toxicology **269**(2–3), 105–119 (2010)
51. De Volder MF, Tawfick, S.H., Baughman, R.H., Hart, A.J.: Carbon nanotubes: present and future commercial applications. Science **339**(6119), 535–539 (2013)
52. Weir, A., Westerhoff, P., Fabricius, L., Hristovski, K., Von Goetz, N.: Titanium dioxide nanoparticles in food and personal care products. Environ. Sci. Technol. **46**(4), 2242–2250 (2012)
53. Sadat-Shojai, M., Atai, M., Nodehi, A., Khanlar, L.N.: Hydroxyapatite nanorods as novel fillers for improving the properties of dental adhesives: synthesis and application. Dent. Mater. **26**(5), 471–482 (2010)
54. Smijs, T.G., Pavel, S.: Titanium dioxide and zinc oxide nanoparticles in sunscreens: focus on their safety and effectiveness. Nanotechnol. Sci. Appl. **4**, 95 (2011)
55. Palaniappan K Is using Nano-silver mattresses/pillows safe? A review of potential health implications on human health due to silver nanoparticles
56. Haefeli, C., Franklin, C., Hardy, K.E.: Plasmid-determined silver resistance in *Pseudomonas stutzeri* isolated from a silver mine. J. Bacteriol. **158**(1), 389–392 (1984)
57. Nair, B., Pradeep, T.: Coalescence of nanoclusters and formation of submicron crystallites assisted by Lactobacillus strains. Cryst. Growth Des. **2**(4), 293–298 (2002)
58. Senapati, S., Ahmad, A., Khan, M.I., Sastry, M., Kumar, R.: Extracellular biosynthesis of bimetallic Au–Ag alloy nanoparticles. Small **1**(5), 517–520 (2005)
59. Jha, A.K., Prasad, K., Prasad, K.: A green low-cost biosynthesis of Sb_2O_3 nanoparticles. Biochem. Eng. J. **43**(3), 303–306 (2009)
60. Li, X., Xu, H., Chen, Z.-S., Chen, G.: Biosynthesis of nanoparticles by microorganisms and their applications. J. Nanomater. (2011)
61. Labrenz, M., Druschel, G.K., Thomsen-Ebert, T., Gilbert, B., Welch, S.A., Kemner, K.M., Logan, G.A., Summons, R.E., De Stasio, G., Bond, P.L.: Formation of sphalerite (ZnS) deposits in natural biofilms of sulfate-reducing bacteria. Science **290**(5497), 1744–1747 (2000)
62. Barthlott, W., Neinhuis, C.: Purity of the sacred lotus, or escape from contamination in biological surfaces. Planta **202**(1), 1–8 (1997)

63. Plattner, L.: Optical properties of the scales of Morpho rhetenor butterflies: theoretical and experimental investigation of the back-scattering of light in the visible spectrum. J. R. Soc. Interface **1**(1), 49–59 (2004)
64. Jeevanandam, J., Barhoum, A., Chan, Y.S., Dufresne, A., Danquah, M.K.: Review on nanoparticles and nanostructured materials: history, sources, toxicity and regulations. Beilstein J. Nanotechnol. **9**, 1050–1074 (2018). https://doi.org/10.3762/bjnano.9.98

An Insight into Properties and Characterization of Nanostructures

Aleena Shoukat, Muhammad Rafique, Asma Ayub, Bakhtawar Razzaq, M. Bilal Tahir, and Muhammad Sagir

Abstract Nanostructures are one of the pivotal structures in nanotechnology that are widely used in various fields like photonics, quantum computation and energy applications. Nanostructures are structures that have size in the range of 1–100 nm. The classification of nanostructures is discussed briefly in the view of confinement of electrons. Based on the quantum confinement of electrons, nanostructures are divided into three categories such as 1D, 2D and 3D nanostructures. Nanostructures possess novel properties such as physical (electronic, optical and thermal), transport and mechanical properties (hardness and fatigue) that are dependent on their sizes. The properties of nanostructures are entirely different from their bulk counterparts because of having large surface-to-volume ratio. It is always difficult to fabricate functional nanostructured surfaces, but the most important activity is the ability to characterize their features, shapes, morphology, size, etc., and then to correlate their physical properties to these characteristics. In nanomaterials, the physical properties include size, morphologies and surface area, and the chemical properties include composition, surface, electrochemistry and oxidation states. This chapter deals with the properties of nanostructures and various techniques required for their characterizations. The main objective of characterization of nanostructures is to achieve a nanoscale comprehension of the physical and chemical aspects of nanostructures. For this purpose, different characterization techniques such as Raman and fluorescence

A. Shoukat · A. Ayub · B. Razzaq
Department of Physics, Faculty of Sciences, University of Gujrat, Gujarat 50700, Pakistan

M. Rafique (✉)
Department of Physics, University of Sahiwal, Sahiwal 57000, Pakistan
e-mail: mrafique@uosahiwal.edu.pk

M. Bilal Tahir
Department of Physics, Khawaja Fareed University of Engineering and Information Technology, Rahim Yar Khan, Pakistan

M. Sagir
Department of Chemical Engineering, Khawaja Fareed University of Engineering and Information Technology, Rahim Yar Khan, Pakistan

© The Author(s), under exclusive license to Springer Nature Singapore Pte Ltd. 2021
M. B. Tahir et al. (eds.), *Nanotechnology*,
https://doi.org/10.1007/978-981-15-9437-3_3

spectroscopy, TEM, SEM, STEM, STM, AFM, molecular modeling, X-ray diffraction, X-ray photoelectron spectroscopy, wide and small-angle X-ray scattering, X-ray tomography and nuclear magnetic resonance are described in detail.

Keywords Nanostructures · Nanowires · Fatigue strength · Surface restructuration · Chemical potential · Hardness · Characterization · Lithography

1 Introduction

The term nano refers to one billionth part of a certain considered unit. Nanostructures are materials that have size range between 1 and 100 nm. Electrons present in nanostructures are confined in one dimension, but free to move in other dimensions [1]. Depending on the confinement of electrons in dimensions, the nanostructures can be classified into quantum wells, quantum wires and quantum dots. The type of nanostructures, in which the electrons are confined in one dimension out of three, is called quantum wells. In quantum dots, the electrons are confined in all the three dimensions of nanostructures. However, in quantum wires, the electrons are allowed to move in one direction only [2].

Nanostructures possess unique microscopic properties as compared to their bulk counterparts. The main difference is related to their sizes. At nanoscale, the properties of nanostructures such as density and conductivity are size dependent, but at bulk level, the properties are independent of their size. The properties of nanostructures are governed by effects such as the confinement of electrons, the location of the surface atoms and quantum coherence. The variation in these effects alters the nanostructure properties [3]. Properties of nanostructures majorly depend on its fabrication techniques, crystalline state and surface–interface effects. Properties of nanostructures can be altered by fluctuating their size [1]. Some large-scale nanostructures arise due to the crystallization or formation of aggregates. Nanostructures with large surface area are of great significance such as polymer blends [4]. Normally, the dimensions of nanostructure materials lie between nano- and micro-dimensions. Nanostructures are coined as the building block of nanomaterials [5].

Nanostructures have great deal of importance in all fields of science. With advancement in nanotechnology and nanostructures, the research field has really exploded with new amusing work [6].

The chapter is divided into two sections. The first section gives insight about nanostructures, their basic classification and fundamental properties. The second half discusses the fundamental characterization techniques.

1.1 Types of Nanostructures

Nanostructure materials are composed of structural elements that have dimensions from a few nanometers to several hundreds of nanometers. The structural elements of nanostructures have a very long range and thus are distorted, the range of multi-particles that have short range governs the arrangement of atoms and molecules in nanostructures [7]. The properties of nanostructures depend on the size and arrangement of their structural elements. The nanostructures can be classified as quantum well, quantum wires and quantum dots on the basis of confinement of electrons, as discussed earlier [8].

Quantum wells are the basic nanostructures in which electrons are free to move in any two dimensions and are confined in the third dimension. Quantum wells are two-dimensional systems that can be realized by squashing a low bandgap material between two high bandgap materials. On the other hand, quantum wires include nanowires and nanotubes, while quantum dots include the nanoclusters [9].

The nanostructures can also be classified as zero-dimensional (0D), one-dimensional (1D), two-dimensional (2D) and three-dimensional (3D) nanostructures according to their dimensions as discussed below.

1.1.1 1D Nanostructures

One-dimensional nanostructures are widely studied to understand the size- and dimension-dependent properties of nanostructures such as transport, optical and mechanical properties. 1D nanostructures are fundamental for the applications in electronic and optoelectronic devices because they are utilized as functional components and to connect components. One-dimensional nanostructures are of prime importance in research and technological applications. One-dimensional nanostructures such as carbon nanotubes and quantum wires are ideal to study the size and dimension dependence of their transport and mechanical properties [10].

1D nanostructures have attained attention due to their application in electronic devices and their one-dimensional morphology such as nanowires. Practical applications of 1D nanostructure utilize their 1D morphology, their defect tolerant character and stability [11]. 1D nanostructures such as nanowires are used in solar cells and batteries. Different morphologies of 1D nanostructures and their respective applications are described in Fig. 1 [12].

1D nanostructures have astounding properties such as high luminescence efficiency, low threshold for lasing action, enhanced figure of merit and mechanical toughness that make them attractive for various applications. Nanowires are anisotropic materials having large aspect ratio. Figure 2 presents different InGaP nanowires at various temperatures [13].

Generally, the diameter and length of nanowires are in the range of several nanometers and micrometers, respectively. Nanowires have different morphologies

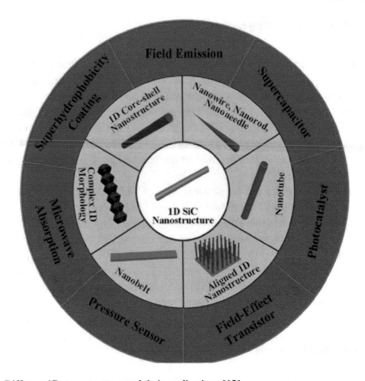

Fig. 1 Different 1D nanostructure and their applications [12]

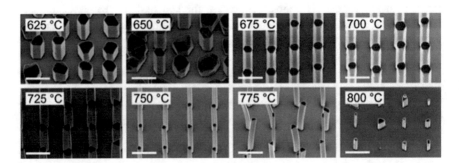

Fig. 2 InGaP nanowires at various temperatures [13]

and properties as compared to spherical nanocrystals. The assembly of 1D nanostructures is an important factor in their practical applications. 1D nanostructures can be fabricated through various techniques such as X-ray lithography, nanolithography, e-beam writing and proximal probe patterning. Figure 3 shows several techniques used for the fabrication nanostructure at different scales.

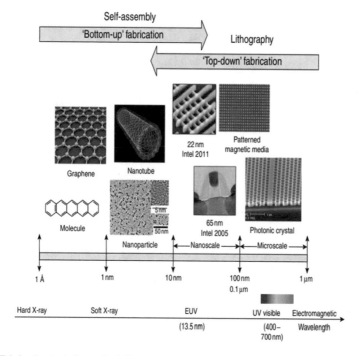

Fig. 3 Fabrication techniques for 1-D nanostructures [14]

1.1.2 2D Nanostructures

2D nanostructures are the materials in which two dimensions are out of the nanoscale. Over the past few years, the research has been focused on the fabrication of two-dimensional nanostructured materials. 2D nanostructures are being utilized in various fields such as photocatalysis, nanosensors and nanoreactors because of having novel properties and small size. The classification of nanostructures and their examples is shown in Fig. 4. The examples of 2D nanostructures are nanofilms, nanolayers, nanoplates, nanoprisms and nanosheets, etc. [13].

Figure 5 shows SEM images of different kinds of 2D nanostructures. The 2D nanostructures include junctions, branched [15], plates [16], sheets [17], walls [18] and disks [19], etc.

1.1.3 3D Nanostructures

The fabrication of three-dimensional nanostructures has been garnered much attention, since the past few years, due to their astonishing properties like large surface area and quantum size effects as compared to their counterpart bulk properties. The performance and potential applications of 3D nanostructures depend on their sizes,

0-D	1-D	2-D	3-D
Nanoparticles, quantum dots 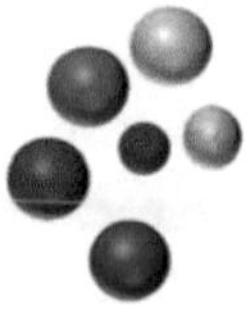	Nanotubes, nanorods, nanowires 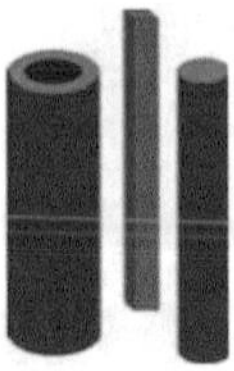	Nanofilms, nanolayers, nanocoatings	Bundles of nanotubes, nanowires, multi-layered materials 

Fig. 4 Classification of nanostructures [13]

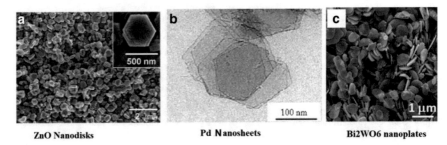

ZnO Nanodisks   Pd Nanosheets   Bi2WO6 nanoplates

Fig. 5 SEM images of different morphologies of 2D nanostructures **a** disks [19], **b** sheets [17], **c** plate [16]

particular morphologies and dimensions. The controlled fabrication of 3D nanostructures for different morphologies is of prime importance for the practical usage. 3D nanostructures are widely employed as catalysts and as electrode or magnetic materials. 3D nanostructures have also attracted the attention of researchers because of having more absorption sites. 3D nanostructures include nanocoils, nanopillars or nanocubes [19].

2 Fundamental Properties of Nanostructures

Nanostructures are called mesoscopic structures having unique properties that are different from their behavior at bulk level or at atomic scale. The focus of nanoscience is the unique properties of nanostructures. New and advanced chemical, mechanical and optical properties of nanostructures are possible through their controlled fabrication. The properties of nanostructures depend on surface effects and their size at nanoscale. Comprehension of different properties of nanostructures and their characterization are important factors to understand the structures of different nanostructures and their function. The atomic configuration of nanostructures depends on the kinetics of its fabrication methods. It is possible to control the properties of nanostructures by adjusting the parameters such as concentration, temperature and pressure during fabrication. Nanostructures have better performance due to their large surface-area-to-volume ratio. In the following sections, some fundamental properties of nanostructures followed by their characterization methods are discussed [20].

2.1 Physical Properties of Nanostructures

Nanostructures have different properties from their bulk material due to their large surface area, confinement of electrons and other quantum effects. Nanostructures possess specific electric, optical, chemical and thermal properties.

2.1.1 Thermal Properties

Thermal stability is one of the important parameters of nanostructures that govern their application in nanoelectronics. Nanostructures possess different morphology and microstructures depending upon temperature such as silicon nanostructures. Formation of nanostructures is dependent upon their annealing temperature. By controlling the annealing temperature, one can control and adjust the morphology and structure of silicon-based nanostructures [21].

According to the second law of thermodynamics, entropy rate is positive for a system that is large with longer timescale. But smaller systems with smaller timescale violate this law according to the fluctuation theorem, and this has been proved experimentally. The experiment involves the observation of the position and forces acting on a particle that is located inside an optical trap. Analyzing the entropy is important for application of nanostructures in nano-machines such as nanomotors. Probability of thermodynamically running of nano-machines increases with miniaturization of these nano-machines [22].

The heat flow or transport in nanostructures is different from the classical picture. The flow of electrons that carry the heat depends upon the gate voltage. Phase transition in nanostructures is different from bulk materials and depends on particle size, interaction between nanostructures and its substrate and wetting [23].

2.1.2 Optical Properties

Nanostructures like nanowires possess optical properties such as absorption and luminescence due to quantum confinement effects. Silicon nanowires exhibit a blue shift at absorption edge. Discrete absorption properties and photoluminescence at the band edge are also exhibited by silicon nanowires due to quantum confinement and surface effects [24].

Confinement of electrons in 0D nanostructures and 2D nanostructures leads to the variation in band structure. The variation in band structure thus leads to variation in optical properties such as emission and absorption energies. To understand the confinement of electrons and their effect on optical properties, Schrodinger's equation is solved with some assumptions for bound state energies and wave functions. Nanowires exhibit lasing action at room temperature in an ultraviolet region. Nanostructures, with smaller diameter, do not exhibit lasing action due to diffraction and photoluminescence effects. Nanostructures having high nonlinear optical response can be employed for nanoscale optical circuits. Nanostructures of wide band gaps are used in solar cells [25].

2.1.3   Electronic Properties

Semiconductor nanowires are employed in several electronic devices such as BJTs, FETs, PN diodes and inverters. Electronic properties of semiconductor nanowires can be controlled during their assembly and synthesis. Bottom-up fabrication of nanostructures is better than top-down fabrication techniques. Electron transport properties become significant as the size of nanostructure decreases. The density of nanodevice increases by decreasing their size [26].

Confinement of electrons in one, two or three dimensions leads to smaller scattering of charge carriers. The reduction in carrier scattering is the decrease in accessible k-points due to confinement. The mobility of electronic carriers is affected by scattering in multiple ways as scattering with other carriers, surfaces, interfaces, phonons, impurities or Plasmon's.

Electronic properties of nanostructures act as bridge between bulk and molecular materials [26].

Dimensionality and size greatly affect electron transport in nanostructures. Nanostructures with large diameter and surface-to-volume ratio affect electron transport properties. The variations in electron transport properties are dependent upon the confinement of excitons and carriers in nanostructures [27].

2.2   Mechanical Properties of Nanostructures

Mechanical properties of nanostructures are fundamental for their application in microelectromechanical devices (MEMS). MEMS refer to a system in which length is between one millimeter and one micrometer. Deformation properties under elastic and inelastic regime for nanostructure are important to characterize their strength. The structural strength is very important in device operation. Some of the mechanical properties include hardness, elastic modulus, fracture toughness and fatigue strength. Bending strength of nanostructures is another important mechanical property of nanostructures. AFM is used to determine mechanical properties of nanostructures [28, 29].

Grain size plays the most dominant role in mechanical and functional properties of nanostructures. Strength of nanostructures is inversely proportional to the grain size as the grain size reduces by increasing the strength [30].

Some fundamental mechanical properties such as hardness, fatigue and bending strength are briefly discussed in the next sections.

2.2.1 Hardness

Hardness is uniform for nanostructures when the temperature for testing hardness is smaller than the consolidation temperature. But with the increase in consolidation temperature, flaws can be increased in nanostructure that further changes the grain size [31]. In bulk materials, hardness does not depend on grain size, but at nanoscale properties, it becomes size dependent. Hardness increases as the grain size decreases at nanoscale [32]. Hardness decreases at very high temperatures due to creep flow. Hardness of nanostructures is dependent upon the grain size for particular range of temperature. Hardness of nanomaterials is about five times greater than their bulk counterparts. According to the Hall–Petch effect, the increase in the grain size might decrease the flaws in nanostructure and varies the internal stress at very high consolidation temperature. According to the inverse Hall–Petch effect, hardness increases with the increase in grain size [31]. An ambiguity still exists in the literature about the authenticity of inverse Hall–Petch effect. Hardness and elastic modulus are calculated from load displacement data recorded by nanoindenter monitor. Higher the hardness, higher the load will be required to achieve the indentation depth [33].

2.2.2 Bending Strength

Bending tests are used to calculate the elastic modulus of nanostructures along with bending strength that is also called as fracture stress. Atomic force microscopy is used to get a topological image of nanostructure by applying a controlled force. Young modulus can be expressed as follows by using beam bending Eq. 1.

$$E = \frac{l^3}{192I}m \tag{1}$$

where l is the length of nanowire, I is the cross-sectional moment of inertia of the beam and m is the gradient of the force versus displacement curve during the bending of specimen. The beam bending equations assume that beams observe linear elastic theory [34].

Some mechanical properties like hardness and elastic modulus of nanostructures are given in Table 1 [28].

Table 1 Mechanical properties of different nanostructures [28]

Nanostructures	Hardness (GPa)	Elastic modulus (GPa)	Fracture toughness (MPa $m^{1/2}$)
Undoped Si (100)	12	165	0.75
Undoped poly silicon film	12	167	0.11
SiO_2	9.5	144	0.85 in bulk
SiC film	24.5	395	0.78
Ni–P film	6.5	130	–
Au film	4	72	–

2.2.3 Fracture Toughness

Atomic force microscopy is used to measure fracture toughness by using indentation technique. Indentation technique involves the measurement of radial cracks produced. Optical microscopes are used to observe the cracks. The fracture toughness is measured by a relation described in Eq. 2.

$$K_{1c} = \alpha \left(\frac{E}{H} \right)^{1/2} \left(\frac{P}{c^{3/2}} \right) \tag{2}$$

where H is hardness, E is elastic moduli as described in Eq. 2, P is the peak indentation, α is constant and c is crack length. Optical microscope is used to measure the length of crack from its center to its rim [35].

Different methods are used to measure fracture toughness such as four-point Single Edge Notch Bend (SENB) specimen technique [36]. To measure fracture toughness, nanostructures should have uniform stress during bending tests [37]. Fracture toughness of different nanostructures is given in Table 1. It is clear from Table 1 that silicon carbide film has higher fatigue toughness at nanoscale as compared to its counterpart bulk material.

2.3 Physical Chemistry of Nanostructures

Nanostructures have large surface-to-volume ratio. The number of atoms present at surface varies from inner number of atoms. Due to this dramatic change in the number of atoms in nanostructures, the variation in grain size greatly affects its physical and chemical properties. In the following section, we discuss how the grain size and other factors affect the physical chemistry of nanostructures [38].

2.3.1 Surface Energy

Surface energy of nanostructures is reliant upon the dimension-dependent surface area. Surface energy increases by increasing the surface area. Surface area and surface energy become negligible at larger scale, but it only becomes significant when material is cut into smaller scale. Surface area and surface energy can be increased seven times by reducing the size of the particles up to nanometers. [39].

Nanostructures are usually unstable or metastable due to their large surface energies. One of the important factors in fabrication of nanostructures is to control their surface energy. To produce stable nanostructures, understanding of the surface energy is very important.

On surface of nanostructures, atoms have incomplete bonds due to their low coordination number. Due to these incomplete bonds, an inward force acts on the surface atoms and thus reduces their bond length [40].

At nanoscale, the decrease in bond length reduces the lattice parameter. Surface atoms possess extra amount of energy that is represented in Eq. 3.

$$\gamma = \left(\frac{\partial G}{\partial A} \right)_{ni,T,P} \tag{3}$$

As discussed earlier that surface energy has an important role during fabrication of nanostructures. According to thermodynamics, any system is stable when it has low Gibbs free energy. To make stable nanostructures, reduction of surface energy is of prime importance. When surface area is fixed, surface energy can be halved via various factors such as surface relaxation, surface restructuration of incomplete bonds and surface adsorption. In surface relaxation, the surface atoms are driven inward rapidly. In surface restructuration, the incomplete bonds at surface are converted into bound chemical bonds. Surface adsorption involves the introduction of chemical species on the surface through physical or chemical adsorption. Another process to reduce surface energy is composition segregation that includes the enhancement of impurities on surface through diffusion. The mechanisms of surface relaxation, surface restructuration and surface adsorption are explained in the following sections [38].

2.3.2 Chemical Potential

Chemical potential is a property that depends on surface radius of curvature. Chemical potential can be written as in Eq. 4.

$$\Delta\mu = \frac{2\gamma\Omega}{R} \tag{4}$$

The above expression describes that chemical potential is inversely proportional to the radius of curvature R. Chemical potential is higher for curved convex surfaces as compared to flat surfaces, because curved convex surfaces have positive radius R. Convex surfaces have high chemical potential as compared to concave surfaces [41].

When two particles having different radii of curvatures are dissolved with a solvent, both particles will achieve equilibrium via diffusion from small particles to large particles. Solute particles will diffuse to the area of large particles [42]. This process is called Ostwald ripening. Each particle will possess different chemical potentials.

Ostwald ripening can either increases or decreases the grain size distribution. Abnormal grain growth due to Ostwald ripening can lead to unwanted results such as degradation in mechanical properties of nanostructures.

Ostwald ripening is utilized to decrease the size of nanostructures. The main force behind Ostwald ripening and surface restructuration is reduction in surface energy. Reduction in grain size will reduce surface energy. To have desired nanostructures during fabrication, reduction of surface energy is very important. Another important factor to achieve stable nanostructures is to prevent agglomeration through electrostatic stabilization [43].

2.4 Transport Properties of Nanostructures

Transport properties require proper theoretical understanding of nanostructures. Different models have been studied to understand transport properties of nanostructures. Transport properties of nanostructures are related to the size of the system and its length characteristics. The length characteristics of nanostructures depend upon temperature, electric and magnetic field and impurity scattering. Transport properties vary from material to material.

The momentum of electron in nanostructures is disrupted due to scattering. One of the important scattering mechanisms that affect the transport of electron is the impurity scattering. The mean free length is directly proportional to relaxation time of momentum. It is the distance traveled by electron before which its coherent state is disrupted by inelastic scattering. The phase of electron is disturbed by scattering such as due to electron–electron scattering or electron–phonon scattering. Impurity scattering can also impart an effect on phase relaxation length. If impurities have internal degree of freedom such as internal spin in case of magnetic impurities, then it can change the phase of electron and hence phase relaxation length.

Kinetic energy of electron is dependent on de-Broglie wavelength, which determines the point on the length scale at which the wave-like behavior of electron becomes important. It is associated with the energy of thermal electrons that is kT. Thermal fluctuations make the phase of electron traveling at Fermi velocity, i.e., unrecognizable within thermal length [44].

2.5 Physical and Chemical Properties of Nanostructures

Though nanostructures have the chemical composition identical to that of a single atom or bulk material, there exist variations in their physical and chemical properties. The shape, spatial or electronic structures, energies, reactivity, catalytic properties, their assemblies and phases are related to their molecular and nanostructure. Size, shape, electric response and quantum confinement are crucial features of nanostructures. Size variation revolutionizes the thermodynamic, electronic, spectroscopic, structural and chemical behavior of these systems. In addition, morphology has exceptional effect on the chemical and physical properties of nanostructures. The shape of the nanostructure is also correlated with their properties as well as their molecular interaction in contrast with the bulk materials. In case of nanocomposites, not only morphology, but the nature of interaction also becomes prominent. 1D confinement effect along with the size effect of individual nanomaterials arises in the case of multilayered structures and thin films. Similarly, in multilayered structure, 1D periodicity arises due to atomic arrangement and interfaces of the materials, which in turn modifies the physical characteristics of the material [45].

Single or multiphase polycrystals correspond to the nanostructures because of their microstructural features, domain size and thickness in nanorange. Attributable to the lowest dimensions, nanostructures have considerable atomic fraction in grain or interface boundaries and thus possess unique physical and chemical properties. To enhance the properties and surface area of material, the pore sizes are reduced in the nanoregime. The exploration of size and shape can be illuminated by Fig. 6, where variations in size and shape of nanostructure modify the Van der Waals forces. The stability, utility and reactivity of the material are dependent upon the chemical behavior of the interfacial atoms. The higher surface area of the nanostructures is achieved by the fabrication of smaller clusters or particles. Zeolites, amorphous silica and porous carbon are some of the examples of nanoporous materials [5, 46].

2.6 Distinctive Properties of Nanostructures

Nanostructures possess unique and distinctive properties like ultrahigh mechanical strength and high electrical conductivity. Unique optical behavior such as blinking is shown by CdSe quantum dots. The quantum behavior becomes more dominant by reducing the size. The nanostructures display classical and quantum blend behaviors and offer new promising technologies. Semiconductor quantum dots show fluorescent behavior, which is explained by means of quantum mechanics that the tunneling current flows by the emission of electrons form the tips of Bucky tubes. The giant magnetic resonance (GMR) materials show electrical resistance in response to magnetic field and are widely employed in storage devices to store magnetic information. Similarly, magnetic semiconductors display behavior of spin polarization of

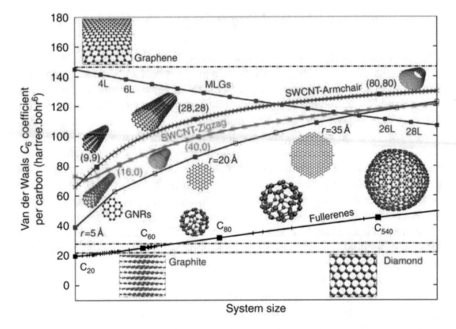

Fig. 6 Scaling law of quantum coefficients [47]

electrons and form the basis of spintronics. Nanosensors can be formed by fabricating tips, wires and small cantilevers [29].

3   Characterization of Nanostructures

Different characterization techniques are utilized to explore several properties of nanostructured materials such as structural, morphological, optical, electronic and magnetic. Nanocrystalline materials are usually characterized on atomic and nanoscale to observe their properties. The interactions of nanoparticles with other particles must be considered to characterize these nanostructures [48].

3.1   Materials Characterization

The characterization is a process through which one can explain the basic attributes of the material under study. The material can be characterized at the applied or development stage and ensures consistency, reliability and process validation. Every material has unique attributes that can be observed by using various characterization techniques. At nanoscale, characteristics and behavior of material become size-dependent, and thus, the quantum effects dominate. [49].

3.2 Photon-Optical Spectroscopy—Raman and Fluorescence

For nanostructure characterization, extensively implemented techniques are Raman and fluorescence spectroscopies. Though they have some incompatible conditions like spatial resolution is limited by diffraction in both techniques as well as their resolution limit is up to 0.25 μm, still Raman spectroscopy is mostly utilized in case of carbon-based nanostructures. Here, it couples with lattice or intramolecular vibrational modes of structures within crystalline order. While in case of quantum well structure, fluorescent spectroscopy is preferred, and this involves coupling of atom-like electronic states. Marginal preparation of sample is required in both cases. Qualitative and semi-quantitative evidence can be concluded by analyzing the unique spectra formed in the spectroscopy [49].

3.2.1 Raman Spectroscopy

Material scatters incident photon from its surface after the illumination of light. In case of interaction with original frequency, photons are scattered and give rise to elastic or Rayleigh scattering. While in case of loss of energy during photon scattering, the entire process is named as Raman scattering. Variation in the frequency or wavelength of the photons gives rise to higher or lower stokes in the spectrum due to coupling of photons with electronic and vibrational molecular states [49].

In a conventional Raman spectroscopic process, one out of 107 photons participates in Raman shifting and thus makes the process inefficient. This efficiency is increased by introducing different absorbents, like gold or silver, on the nanostructured metal substrate, and the phenomenon is regarded as surface-enhanced Raman spectroscopy (SERS). In this process, incident photons excite the surface plasmon of the metal, by equalizing their frequencies. This results in the creation of strong near-field radiation, which immerses all the nanostructured materials or adsorbed molecules. SERS in combination with probe instrumentation is considered as the most reliable technique because they can form spectra from a single nanostructure or a solo molecule [50]. When SERS is combined with a sharp metallic tip or fiber-optic probe, the arrangement is named as tip-enhanced Raman spectroscopy (TERS). This technique involves the generation of near-field specimen illumination through stimulating plasmon on surface of the tip. SERS involves absorption of specimen in the metallic substrate, while TERS involves generation of spatially resolved spectrum through single molecule [51].

The lateral resolution of Raman spectroscopy can be improved by means of confocal arrangement in addition to in-depth resolution so that it can transfer the

Fig. 7 Schematic diagram
of Raman confocal
spectroscopy [52]

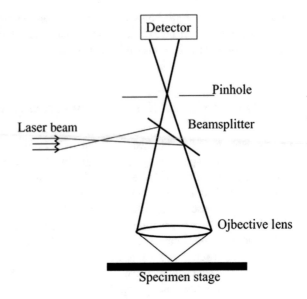

interaction volume into the specimen bulk from that of surface and thus results in-depth profile. Motorized x-y-z sample stage can pave the way for 3-D plotting of the specimen. Schematic diagram of Raman confocal spectroscopy is displayed in the Fig. 7.

Characterization of nanosized structures and materials is done via Raman spectroscopy by taking into account various factors such as the phonon confinement, strain, temperature effects, lattice distortion, grain size, substitutional and structural defects. These factors widely affect position of peaks, modes, shapes and line width of Raman spectra of nanostructures. Nano-powdered oxides like TiO_2, CeO_2, ZnO can be analyzed by Raman spectroscopy in addition to multilayered nanostructures like $ZnSe/SiO_x$. It basically involves the comparison of broadening and shifting of Raman modes obtained by nanostructures to that of the spectra formed by phenomenological modes considering harmonicas and disorder effects through breakdown at $k = 0$ via 3 to 4 phonon decay process. Raman peaks of CeO_2 show strong confinement and inhomogeneous strain, while TiO_2 and ZnO display dominancy of harmonic effects and tensile strain. In case of $ZnSe/SiO_x$ multilayered structure, shifting and asymmetric broadening of modes are observed via one-dimensional confinement model, which is mostly implemented in case of quantum well structure or thin films. When nanostructures are analyzed via IR spectra, it is relatively distinctive than ordinary nanocrystals because of polycrystalline nature and structuralism. All information about size, gap, porous nature, surface bonding and chemical reactivity can be obtained by mean of IR spectra [48].

3.2.2 Fluorescence Spectroscopy

In the fluorescence process, a photon is absorbed to create excitation of atoms from ground state to an excited state by electronic transition and thus introduces vibrational modification. By achieving the lowest vibrational states while relaxing in the excited state, there comes relaxation to other vibrational states along with the emission of fluorescence photon with certain energy. Fluorescence energy spectrum is formed due to the presence of several modes of vibrational states destination. With the help of fluorescence spectroscopy, electronic as well as vibrational structure of the specimen can be visualized. Thus, it is more efficient than Raman spectroscopy. To detect single nanostructure through fluorescent technique utilizing optical probe, there are certain conditions that must be met. This includes interaction of the probe volume with only one nanostructure at a time, and the structure should be fully dispersed on the substrate or in fluid form. The radiations scattered from the nanostructure should be analyzed by the chosen spectroscopic technique. Moreover, optical path components should not interrupt in the fluorescence [53].

Carbon nanotubes (CNTs) can be analyzed by fluorescent spectroscopy, and the schematic diagram for SWCNT is presented in Fig. 8. Illumination from the lamp is obstructed and sample of SWCNT is excited via diode laser at three variant wavelengths. Near-IR emission spectra are obtained at 180° and then investigated by multi-channel array. Tungsten laps are unblocked, and lasers are shut down to obtain absorption spectrum. The near-IR light spectrum through SWCNT is detected by the system. Absorption spectrum is then obtained by using a blank sample for transmission, and then, both spectra are compared. A programmed table for spectra of

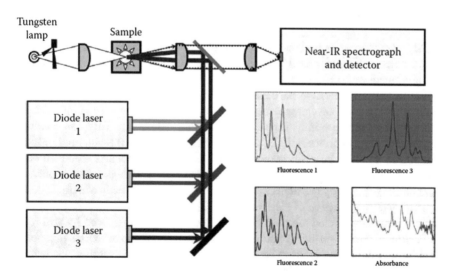

Fig. 8 Schematic diagram of fluorescent spectroscopy for SWCNTs [54]

SWCNT can help in association of each major peak in the emission spectra with the specific species and thus helps in characterizing nanostructures [49].

3.3 Electro-optical Imaging of Nanostructures

The image mechanisms of electron and optical microscopy are similar to each other along with some variations. Diffuse scattering exists in phase contact and transmission in transmission electron microscopy (TEM), while TEM cannot create contrast in backscattering mode. The backscattering image is very popular in scanning electron microscopy (SEM). As the wavelength of electron is comparable to that of interatomic spacing, this leads to the information of spatially resolved structure in diffraction mode. SEM is preferred over routine optical imaging of nanostructures for characterization. It is easy to handle and requires less attention on preparation of specimen, while TEM provides topographical visualization of the nanostructure and promising point-to-point resolution in nanometers due to its current generation techniques. While to make interaction volume reliable with nanoscale examination, SEM is often coupled with EDS. Scanning transmission electron microscopy (STEM) is a little variation in the transmission technique of SEM. A fine beam of electron is focused upon the thin sample in STEM, giving transmission intensity as a function of position. The spatial resolution of STEM is analogous to that of high-resolution TEM (HRTEM) [49].

3.3.1 Transmission Electron Microscopy (TEM)

TEM is considered as a giant optical microscopy with a little variation that high-voltage electron beam is used in it, and lenses are replaced by electromagnetic coils for alignment or focusing of electron beam. These electromagnetic coils perform similar function of bending the electron just like mirrors, and glass lens perform on optical light. As the wavelength of electron is much smaller than that of optical light, this causes an increase in its resolution along with field depth. Very thin specimen usually less than micrometer is required for obtaining the high resolution. For transmission of electron beam inside the instrument, all elements and specimen stage are necessary to be under high vacuum environment. This high vacuum prevents attenuation of beam by gas molecules. Atoms of specimen differ in electronic densities and induce deflection of electron beam. Great disparity occurs because of high atomic number and electron densities. Divergence can be improved by absorbing high atomic number element of choice within a specific phase or region of concern. Osmium tetroxide can be used for the purpose of selective absorption. Schematic diagram of TEM instrument is shown in Fig. 10 (Fig. 9) [55].

Image of sample along with its diffraction can be visualized by TEM technique. It is a leading technique used for nanostructures characterization, attributable to its

Fig. 9 Schematic overview of conventional TEM [52]

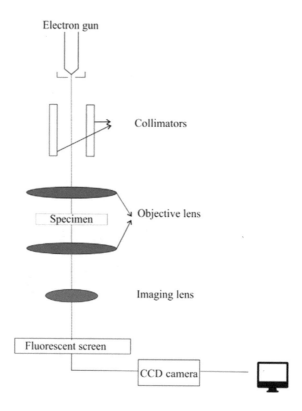

high lateral spatial resolution, which is round about 0.08 nm. Many characterization techniques are unable to explore the particle size, color, elastic and mechanical properties of the individual particle, while modern TEM is developed with additional features that they can configure nanoelectron size probes in the range of 2–5 nm. A multistage condenser lens system is utilized to make this configuration possible. Scanning transmission is made possible by this lens system, and resolution is enhanced by electron probe diameter. Thus, it has been made possible to form images of thin specimens and samples which possess higher crystallinity along with their thickness consideration. Secondary and backscattered electrons can be recorded by multistage condenser lens system, thus making it more advantageous. Cathode luminescence can have inhomogeneity in it, which is tackled by utilizing complex multistage lens condenser, in which structural defects are considered. Cathode luminescence microscopy technique is a convenient characterization technique for optoelectronics. SEM can study cathode luminescence of nanoparticles and bulk samples, but it is limited by resolution. Therefore, high-resolution microstructure correlation with cathode luminescence is done by utilizing TEM techniques [48, 57].

Due to its high resolution, TEM can be used to characterize polymer blend nanostructures and carbon-based blended materials. There also exist some glitches in TEM that include examination of only small area which cannot describe the whole material.

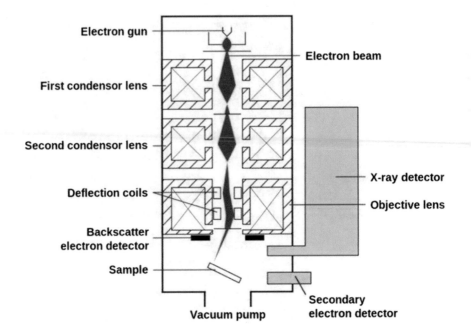

Fig. 10 Schematic of typical SEM [56]

Therefore, certain features are inferred and establishing particles network, agglomerates or exfoliation of the material that are usually not demonstrated by the material [55].

3.3.2 Scanning Electron Microscopy (SEM)

Imaging of structures from micro- to nano-regimes dimensions can be formed by SEM. It is similar to TEM as they both utilize electron beam for analyzing the specimen. The way of image formation is different from that of TEM, but the apparatus such as electron gun, evacuated system and condenser lenses are identical in both techniques. It operates in reflection mode and requires less voltage for electron acceleration. Internal structure of specimen is imaged by TEM, while SEM forms the images of upper or near surface and bulk specimen's topography. SEM can work in backscattering mode to expose contrast of atomic number and can be coupled with EDS analytical system at low lateral resolution. These two imaging modes are inappropriate in case of nanostructure characterization, because of high interaction volume. In SEM, electromagnetic optics appear in the incident probe beam, which is then focused on surface of specimen with the help of condenser lens and objectives. Spot size, diameter of electron source and optical deviations significantly influence point-to-point resolution [49]. Schematic diagram of typical SEM is displayed in Fig. 10.

Resolution of SEM is up to 1 nm, and its magnification is about 400,000×. In SEM, preparation of specimen is relatively easier, and it allows direct observation of few materials. For insulating material, a conductive coating is applied on it for conduction of current from beam of electron to stage. Image formed by SEM can expose surface contours of the material and thus provide all information about material, its preparation and impairment. SEM can also implement extracting or surface etching techniques to discriminate surface components and thus disclose nature of the material under study.

To compensate voltage requirement and vacuumed conditions, environmental SEM (ESEM) is designed which can operate at low voltage and in low vacuum. Materials with bits of moisture like protein, cellulose and fats can be analyzed by means of ESEM, where image of specimen is formed by secondary electron signal. For detection of specific elements and their quantity, X-ray analysis is done by providing high-voltage beam of electron. This is known as energy-dispersive X-ray (EDX) technique, which is utilized in both TEM and SEM [55].

A field emission cathode is implemented in SEM electron gun to provide thin probe beam, through which spatial resolution can be improved along with fewer sample charging. This is regarded as field emission SEM (FESEM). In this technique, electron gun is designed differently to increase focusing of electrons. Very high electric field is applied very near to the tip of filament, and electrons are ejected out of it. FESEM is of prime importance for nanostructures characterization either in the form of nanosheets, nanoparticles, nanoflowers or thin films. When molybdenum disulfide nanosheets are characterized by FESEM, they display striking conductive, magnetic, photocatalytic and FET properties. Properties shown by nanostructures are greatly determined by methods of their generation and structural characteristics. Nanosheets have been used in various applications such as energy harvesting, optoelectronics and spin electronics [48].

3.3.3 Scanning Transmission Electron Microscopy

Scanning transmission electron microscopy (STEM) technique is the extension of both TEM and SEM techniques. Electron beams are scanned across the specimen, thus making the STEM suitable for EDX analysis, annular dark-field (ADF) imaging and electron energy loss spectroscopy [58]. This leads to form an association between quantitative data and image formed. An atomic resolution image can be formed by STEM utilizing high angle detector, where atomic number effect contrast is regarded as Z-contrast images [55].

3.4 Scanning Probe Techniques and Methods

Scanning probe microscopy (SPM) is a surface characterization technique through which topography of nanostructures can be explored along with other properties. The

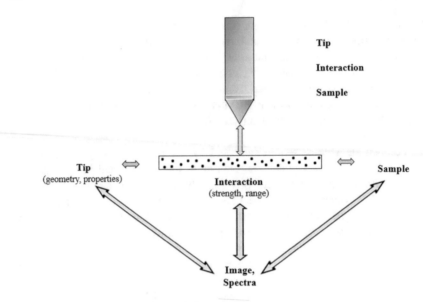

Fig. 11 Essential elements of SPM technique

main members of SPM family are scanning tunneling microscopy (STM) and atomic force microscopy (AFM). STM and AFM were developed by Rohrer and Binnig in 1982 and by Binnig et al. in 1986, respectively. It basically consists of physical probe or tip with apex in nanometer dimensions which is passed over the surface with the help of piezoelectric actuators, thus imitating structures of the surface of material under study. These techniques differ in surface to tip interaction due to various mechanical, electric, optic, thermal or magnetic effects [48]. The schematic diagram of SPM is displayed in Fig. 11. This allows two-way data transmission among the tip and surface [49].

Some other members of SPM family are scanning thermal microscopy (SThM), scanning tunneling microscopy inverse photoemission (STMIP), scanning tunneling spectroscopy (STS), scanning ion current microscopy (SICM), scanning near-field optical microscopy (SNOM), scanning capacitance microscopy (SCM) and spreading electrical current microscopy (SCEM). Their striking features and schematic description are shown in Fig. 12 [49]. A local sensor based on thermal radiations mainly thermistor or thermocouple is utilized as a tip in case of SThM, while, in case of STMIP, low-energy photons are emitted in response of the tunneling current which flows due to injection of electrons in unoccupied surface states. Ion current drawn via a nanocapillary tube in the vicinity of the surface is identified in SICM technique. All other techniques are based upon scientific discoveries and new technologies. Likewise, SPM was aroused due to precursor technology along with local probe techniques, thus considering as a vital technique for surface characterization [49].

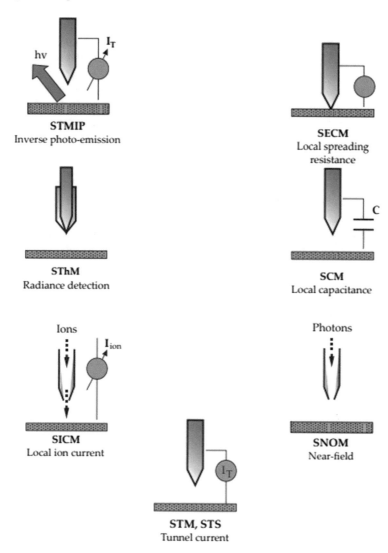

Fig. 12 Schematic illustration of various SPM techniques

3.4.1 Scanning Tunneling Microscope

STM technique is one of the SPM techniques which form images of surface with greater atomic accuracy. A tip, mainly formed of platinum, tungsten, gold or carbon nanotubes, is moved across the material's surface. The direction of tip is controlled by a piezoelectric actuator. High-resolution images of conductive or semiconducting surfaces can be formed by this technique, with a lateral resolution of 0.1–0.01 nm in depth. For this resolution, vibration of sample must be controlled while measuring, thus mostly measured in vacuum environment. Tip is brought near the surface to an equilibrium position usually 4–7 Å to establish attractive or repulsive force. With further decrease in distance, electrons begin to tunnel among the tip and surface, providing tunneling current. With the help of this current, surface height can be determined. It can be operated in two modes, either constant current or constant height mode. In these modes, current and height are made constant, and surface is profiled. Latter mode is faster as time required to keep voltage constant via electronic control is less [59, 60].

Suitable sample treatment and preparation processes are essential for STM analysis. STM can form images of organic molecules which are placed on conducting substrates along with the availability of ultra-high vacuum (UHV). STM has analyzed the formation of Pd catalyst nanoparticles on the TiO_2 substrate. Tungsten trioxide (WO_3) thin films have been analyzed through STM by Maffeis et al. under UHV within temperature range of 100–900 °C. STM images in Fig. 13 show amorphous particles in 35 nm range under 600 °C, while crystallization of nanoparticles is above this limit [59].

3.4.2 Atomic Force Microscopy (AFM)

Atomic force microscopy or surface force microscopy is considered as a powerful tool for characterization of NMs or NPs [62]. It has gained attention of many researchers due to its flexibility as well as good performance in various modes to study different types of surfaces and mediums such as vacuum and air. It is a technique which is used to get information about physical properties including, roughness, surface texturization, morphology, friction, magnetic and elastic properties, etc. [63–67]. In AFM, a sharp tip is used having diameter of about 10–20 nm [68] which is connected to 100–400 μm [69] long cantilever. The tip and cantilever are fabricated at small scale by using Si or Si_3N_4 [68], since these materials permit moderately modest large-scale manufacturing utilizing semiconductor innovation. In this, tip probe is moved over the surface specimen in response to the tip–surface interaction. This movement is measured by focusing laser beam with a photodiode. A systematic setup is represented in Fig. 14 [69].

The geometry of apex of tip plays a vital role in the determination of resolution. The apex of tip senses "individual atoms" on surface of the specimen due to formation of incipient chemical bonds with each atom. These chemical interactions precisely alter the tip's vibrations which can be measured and mapped. The cantilever along AFM tip has three-dimensional motion relative to the specimen surface which is obtained by piezoelectric crystal. The given setup allows us to position the AFM tip more precisely along x, y and z directions. When AFM tip and surface are in contact, the cantilever bends as a result of force between tip and surface of sample. This deflection is utilized for visualization of the surface topography or as input source for the feedback loop (FL). This feedback loop controls the z-position of AFM. There are number of techniques that can be used for estimating deflection, especially an

Fig. 13 STM images of WO$_3$ thin films at **a** 300 °C, **b** 600 °C, **c** 700 °C, **d** 800 °C at constant current mode [61]

optical sensor (OS). In OS, laser light is reflected from the cantilever's back side onto light detector which is sensitive to position [69].

An AFM can be operative in two basic modes; one is contact mode, while other is trapping mode. In contact mode, tip has continuous interaction with the surface, while in trapping mode, cantilever vibrates to and fro vertically and tip has a periodic contact with the surface of the specimen. The contact mode may cause damage to surface. Trapping mode procedure assists with diminishing shear powers is related with the tip development. Trapping mode is mostly used in AFM imaging. The contact mode has limited applications in force curve measurements [70]. Other modes can also be utilized for the enhancement, which have nano-contact between the tip and surface. Specific reagents can be utilized on the probe tip for enhancement, which usually involves biological interactions [55]. AFM has many advantages over SEM/TEM

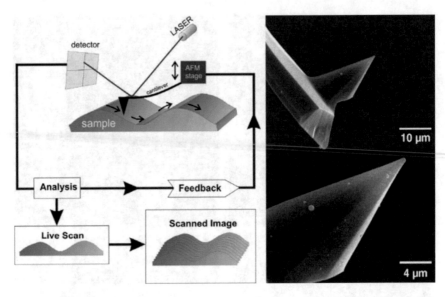

Fig. 14 Systematic setup of surface force microscopy. In this, cantilever/tip is considered as the "heart" of AFM [69]

for characterizing NMs. As image obtained from AFM is three-dimentional quantitatively, therefore, it is easy to measure the height of NPs. In SEM/TEM, usually, a two-dimensional image is obtained. Moreover, AFM is a cost-effective technique for nanomaterial's imaging. AFM requires less laboratory space as compared to TEM/SEM, and as well as it is easy to operate and does not require any trained operator [62]. The resolution obtained from AFM is similar to TEM/SEM, typically "0.2 nm" as well as having a contour height information with precision of 0.01 nm. Advantage of using AFM is that high vacuum is not needed, and it can be operated at atmospheric pressure as humidity or reactive gases. Vacuum is preferred only when there is a need to evade impurities or other fluctuations as a result of surface adsorption or CO_2 [55].The results obtained from AFM are confirmed by utilization of small-angle X-ray scattering (SAXS) [71].

3.5   Image Analysis

After the characterization of material by generating a viewable image, the next step is to quantify image features by extracting numerical data which can be analyzed and correlated with theories. For all imaging techniques, a quantitative analysis of image is essential. Though quantitative analysis of NMs is very important as numerical data's results help us in providing the information about dimensions as well as area of distribution to better look insight the materials. For image analysis, three steps must be conceded.

1. A highly contrast image obtained by different characterization techniques must be converted to a black and white defined field of interest.
2. To recognize the difference between features of interest and background, a threshold must be defined.
3. Different properties of NMs including dimensions, area of cross-section and each particle's distributions must be measured quantitatively.

Different computational softwares developed for image analysis help one to analyze the complex shapes quantitatively. Different properties including aspect ratio, dimensions as well as circumferences can be measured. Image analysis (IA) provides us a significant statistical data which can be compared with other theoretical models, materials, as well as prediction of morphologies with respect to time etc. [4].

3.6 Molecular Modeling (MM)

Molecular modeling is used in variety of theoretical and computational models to study structure of molecules, ions or particles [72]. This method is employed in many fields including quantum chemistry, material science, etc., and ranges from minor chemical systems to biological systems [73, 74]. It usually draws the chemical structures and alters them into three-dimentional NMs. At energy minimization, properties like bond length, dihedral edge as well as bond angles are adjusted depending upon the hybridization of atoms. Energy minimization is done by utilization of quantum, semi-empirical or molecular mechanics to attain a relativistic shape. In this, molecular structures are represented numerically, and then, simulation is done by implementation of classical and quantum mechanics. Furthermore, different properties including bulk, electronic, optical, spectroscopic and geometries are explored [75]. System is studied at molecular level, whereas many molecules are assembled together to get interactions. The simulation of assembled molecules is done by creating a three-dimensional cell in a repetitive order and representing material potion same as that of original material extended within limits. Formally, when model is assembled, energy minimization is done and quantitative measurements are compared with the experimental data acquired from different techniques such as X-ray, SEM and TEM. Simulation is done within a modeled environment or stimuli including pressure, temperature, surface properties, stress, strain and interfacial properties that can be imparted by mathematical modeling. The response of the stimulated model is studied and quantified within imposed environment and then compared with experimental results [55]. The MM limitations include system size as well as time scale which is addressed in modeling. Though, in terms of nanomaterials, NP's size is sufficiently small that can be modeled directly. Different computation softwares are being employed for simulation process. Moreover, simulation's computational costs are a trade of efficiency as well as accuracy [55]. Quantum mechanical-based computation calculations are being employed in the density functional theory (DFT) [76] which is feasible for simulation of few hundred to several thousands of atoms [77] by looking into the determined energies of the template [78]. Additionally, modeling provides a relationship or bridge between experimental and theoretical research and in case of nanosystems, very precise measurements are done theoretically which further can be confirmed experimentally by synthesizing [55].

3.7 X-Ray Based Characterizations

X-ray diffraction (XRD) is a technique which is extensively used for NP's characterization. It is used for exposing information about crystal structure, size of grains, type of crystal structure, as well as lattice parameters. In a specific material, Scherer equation is used for estimation of crystallites size by analyzing diffraction peaks. One of the advantages of XRD technique is the utilization of powdered sample. The particle's compositions as well as peaks intensity of sample are compared with reference data of International Center for Diffraction Data (ICDD). However, for amorphous materials, it is not appropriate, and XRD broader peaks are obtained for particles having size <3 nm [79]. XRD techniques depend upon principle of diffraction, having few basic components as an X-ray source, primary optics, secondary optics and detector. The use of the technique depends upon chemical or physical environment of material to be characterized. In reciprocal space (RS), X-ray pattern is obtained. According to preferred geometry as well as measurement's type, it permits to probe different regions of RS and provides information about the morphologies as well as structures at different length scales [80]. In XRD, different techniques are used which are discussed in the next session of this chapter including small-angle X-ray scattering (SAXS), WAXS, X-ray photoelectron spectroscopy and X-ray tomography.

3.7.1   Wide Angle X-Ray Scattering (WAXS)

The WAXS based on Bragg's law is used to provide link between lattice spacings with angle of scattering of X-ray from the crystal material. In case of NMs, XRD measurements are performed on powdered form of NMs which contains a large amount of random-oriented particles. The X-rays are scattered with the particles and give us subsequent information by showing diffraction rings as well as satisfying the Bragg's equation, and this technique is known as the powder method. In XRD analysis, position and corresponding peak's intensity help us in identification of sample's phase composition. X-ray wavelength has dimensions of order of nanoparticles wavelength. According to Bragg's law, when a beam of light is incident at a particular angle θ on the surface of crystal, n scattering occurs only when beam radiations interact with crystal. The Bragg's condition is mathematically represented as in Eq. 5,

$$2d \sin \theta = n\lambda \tag{5}$$

The diffraction pattern takes place only when the incoming light wavelength is of the order of 1–100 Å of interatomic distance between crystal planes. The X-ray scattering occurs only when incident light angle satisfies the condition, i.e., $\sin\theta = n\lambda/2d$. A part of incoming light deflects at 2θ which leaves behind a reflection spot in diffraction pattern. At all remaining angles, destructive interference occurs. In XRD, scattering peaks represent significant features. The crystal's size is measured by utilizing the Scherer equation which is given in Eq. 6.

$$\tau = \frac{K\lambda}{\beta\cos\theta} \tag{6}$$

Here, τ represents average size which may be less as compared to gain size, K represents dimensionless shape factor, and its value is about unity; λ is incident X-ray wavelength, and β is line broadening representing full-width half-maxima (FWHM), θ represents Bragg's angle [81]. According to the Scherer equation, sharp peaks in XRD represent the high crystallinity of materials [55].

XRD has great potential to analyze NMs, as reflection width and shapes give us great information about the structure of NMs. It deliberates information related to crystal size, dislocation structure as well as lattice's macro-strains, crystal percentage, etc. The crystalline size is the most important factor in powdered diffraction. A diversity technique is utilized for analysis of nano-crystalline powders. Different nanomaterials include ZnO [82–84] and as well as polymer composites are analyzed by utilization of XRD spectroscopy. The polymer composites include polytetrafluoroethylene [85], polycarbonate [86], etc. XRD helps to control the crystallinity of NPs to optimize as well as produce drastic changes in NP's properties.

Figures 15 and 16 show XRD pattern of NiO and $NiCo_2O_4$ microspheres. Nanowires were equipped by utilization of urea-assisted co-precipitation method. XRD pattern is used to identify the crystal as well as purity of the sample. The peaks are diffracted through their respective crystal plane. The average crystallite

Fig. 15 XRD pattern of $NiCo_2O_4$ at annealing temperature of 450 °C as well as at ambient atmosphere [87]

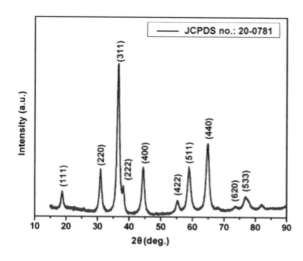

Fig. 16 XRD pattern of NiO
annealing temperature
400 °C [87]

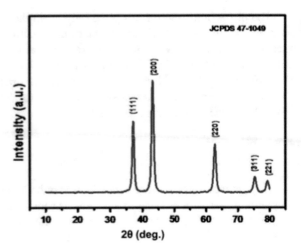

size is obtained by Scherer's equation considering major diffraction peak (200). The
obtained values of K and corresponding X-ray wavelength are 0.89 and respectively.
The average crystallite size was calculated as 45 nm[87].

3.7.2 Small-Angle X-Ray Scattering (SAXS)

SAXS is a small-angle scattering by sample in which X-ray's elastic scattering occurs
whose wavelength is 0.1–0.2 nm. The sample has in-homogeneities in nm range and
therefore observed at very small angles typically 0.1°–10°. The angular arrangement
of SAXS gives us information about pore sizes, shape and size of macromolecules,
etc. In SAXS, to satisfy Bragg's law due to change in d, the value of θ must be
altered. For instance, due to increase in d, corresponding value of θ decreases which
makes it capable of providing information related to crystal structure between 5
and 25 nm macromolecules with partially repeat distances up to 150 nm. Ideally,
SAXS is useful for providing information of ordered system, for instance, nano-
mesoporous constituents, liquid mesophases in crystalline form as well as complex
fluids. In SAXS, X-ray scattering is observed at a very small angle which deliberates
information related to size as well as shape of macromolecules.

The experimental setup of SAXS consists of (1) X-ray source, a pinhole, collima-
tion system having slit, 2D detector and scattering vessel and sample. The sources
of X-rays may be of two types; one is laboratory source such as Kratky or anode
which is rotating (X-rays are produced as a result of charged particles hitting to Cu or
Mo), and the other is synchrotron facility in which charged particles are accelerated
in magnetic field. Slit is collimated due to which higher yields flue, but in this case
slit-smeared data is produced. The Cuk_{α} line is characterized by utilization of X-rays
having wavelength 1.54 Å, and a setup of SAXS is shown in Fig. 17.

Figure 18 represents the results of XRD obtained for TiO_2 nanostructured film.
Each peak represents scattering from a particular crystal plane [88]. SAXS is
significantly used when high electron density of dispersed phase is formed, as it

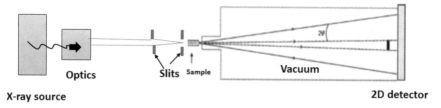

Fig. 17 Experimental setup of SAXS

Fig. 18 XRD pattern of TiO$_2$ nanostructure film [88]

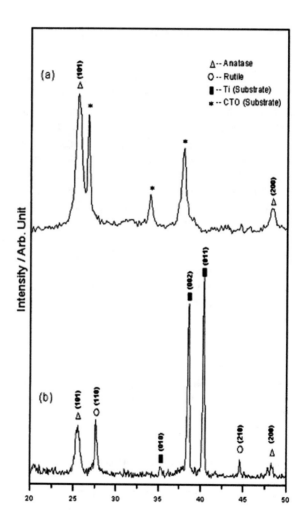

characterizes disperse phases. A group of researchers characterized silver sols by utilization of SAXS, UV–Vis, etc. According to SAXS analysis, a monomodal size distribution of scattering is observed, while according to UV–Vis analysis, a multimodal size distribution of scattering is observed [89]. SAXS technique is used for characterization of different nanomaterials including Ag–Cu alloy NPs [90], polystyrene-*block*-polyisoprene (PS-*b*-PI) [91], polycrystalline, amorphous crystals [55], etc. Ti et al. have written a large review on role of SAXS in characterization of NMs [92].

3.7.3   X-Ray Tomography (Nano-CT)

The tomography based on X-rays is named as CT scan, extensively used in medical field to determine the internal hard structure of tissues. This technique is also utilized for exposing the three-dimensional internal structure along with component distribution of NMs. Three-dimensional imaging data can be obtained by scanning two-dimensional X-ray beam across specimen through rotation of single axis which helps to include three dimensions in finishing image. Due to nano-CT, internal image characteristics can be seen including component distribution of specimen as well as cross sections, defects or plane slice component distribution, etc. CT scan test is used to map volume distribution of material's components in a non-destructive manner. The obtained images give us significant information about image segmentation as well as volume and surface rendering. In case of volume segmentation, a particular threshold is utilized to separate the image components as a result, a particular part of image is omitted. The voxel size of nano-CT is 63 nm [93]. It is used for detection of defects and crystallography at sub-micron resolution [94]. For characterization, typically, two methods are used for stress corrosion cracking detection in nuclear reactor's steels. In this, absorption coefficient's (AC) variation along X-ray beam is found. As AC has direct link with density as well as atomic number of variety of materials on which X-ray beam falls, then a visualization of defects is done easily. Similarly, in nano-CT, a three-dimensional image is gathered by utilization of series of two-dimensional images. Advantage of utilization of this is that it is performed in situ, which gives significant information such as localized corrosion. When 3D characterization of material is done, then X-ray diffraction topography method is considered on top [95–97].

3.7.4   X-Ray Photoelectron Spectroscopy (XPS)

Excess EM energy is transferred to the outermost electron known as Auger electron (AE). AE is analyzed for chemical identification by utilizing the Auger electron spectroscopy (AES). XPS is used for analysis of electron's emission having similar high energy. XPS is used for measurement of different chemical or electronic states of sample's surface. It is based on the principle of photoelectric effect according to which electrons are being emitted on exposure of light [79]. In XPS, the X-rays of aluminum or magnesium are irradiated on the material which is being examined. Production of monochromatic aluminum K_α X-rays from the non-monochromatic X-ray (Non-MCXR) is done by focusing and diffracting these X-rays from the thin crystalline quartz. Such types of X-rays having corresponding values of wavelength, photon energy and energy resolution as 8.3386 Å, 1486.7 eV and 0.25 eV, respectively. Non-MCXR of magnesium having corresponding values of wavelength, photon energy and energy resolution are 9.89 Å, 1253 eV and 0.90 eV, respectively. The emitted electron's kinetic energy is recorded which deliberate the atomic binding energy of the sample's surface elements. A plot between emitted electron's energy and corresponding no. of electrons deliberates a spectrum which provides a qualitative and quantitative analysis of sample NMs composition [98].

Ultra-high voltage is utilized for analysis of XPS, and for this, a typical resolution of 1000 ppm is required. For obtaining this resolution, longer recording times as well as optimized settings are utilized. Due to non-MCXR sources, a significant amount of heat (up to 200 °C) is produced on sample's surface as anode due to which X-rays are produced, only few centimeters away from the surface. This heat causes damage to the sample's surface when joined with highly energized Bremsstrahlung X-rays. Therefore, organic chemicals are not analyzed by utilization of non-MCXR sources [99]. Nag along with his co-researchers published a review unfolding XPS analysis which is a fascinating revenue to investigate internal heterostructure of NPs. For instance, it is used for environmental-dependent crystals as well as tunes the NPs of metal chalcogenides having different sizes [100]. By utilization of XPS, L-cysteine interaction with Au NPs is studied, which experimentally give support to crystal deactivated kinetic model as well as studied that contribution of low coordination Au NPs on corner and edges of NPs. Different nanoparticles including Au@SiO$_2$NPs, Au@Ag core−shell NPs, CdS@Ag$_2$S core/shell NPs, etc., were analyzed by utilization the XPS [101–103].

Figure 19 represents XPS analysis of Sb doped ZnO NPs. According to this, Sb and Zn are found in oxide states. At 530 eV energy, the superposition of $1s$ and $Sb3d_{5/2}$ and direct atomic percentage of Sb by utilization of Sb peak area

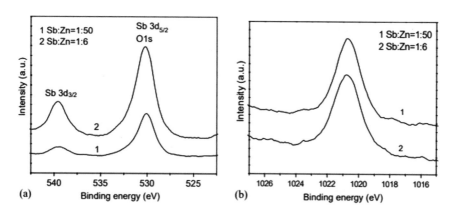

Fig. 19 XPS analysis of **Sb** doped **ZnO** nanoparticles [104]

cannot be determined. However, the ratio of peak areas can be calculated between Sb $3d_{3/2}$ and Zn $2p_{3/2}$, which are 0.4937 and 2.217 at 1:50 and 1:6 correspondingly. It also shows doping content increment with increment in Sb/Zn ratio of evaporated materials [104].

3.8 Nuclear Magnetic Resonance (NMR)

Different techniques such as biophysical and analytical are being employed to study various physical properties of NMs [105–107]. However, only few techniques can give information about nanoparticle attractions. As nanomaterials are exhibiting wide range of morphologies due their surface-to-volume attenuation surfaces, hence it is vital to describe their interactions across the surface to identify attachment sites as well as to quantify interactions strength. Therefore, a very reliable and robust technique is required for characterization of nanomaterials. NMR spectroscopy is considered as an ideal technique which provides unique information about nanomaterial structure's interactions at atomic level without vanishing.

3.8.1 Theoretical Principle of NMR

NMR is an optimistic technique which is performed on the NMs having liquid or solid state to get a deep insight of nanomaterial's structure of molecular frameworks, dynamics, stoichiometry, composition, molecular weight, position and diffusion properties, etc. [108, 109]. Moreover, NMR has an evident cutting-edge feature which helps for characterization of suspensions/colloidal NMs along negligible losses as well as perturbations [58].

NMR is considered as an excellent technique which provides the information of the chemical environment constituents of nuclei without damaging the sample. The principle of NMR is based on nuclei of atoms having magnetic properties, which is further utilized to extract information about chemical environment. Subatomic particles including electrons, protons, etc., have the property of spinning about their own axes. And overall nuclear spin (NS) creates a magnetic dipole along spin axis route. The net spin of the nucleus is "I" having alignment along externally applied magnetic fields (EAMF) by "2I + 1" ways. For instance, if spin of nucleus is ½, then two orientations occur; one is opposing to the external field, while other is reinforcing. In the absence of EAMF, both possible orientations have equal energy. The orientation axis of NS is not oriented with applied field either parallel or antiparallel but must press with respect to applied field at an angle along angular frequency as represented in Fig. 20 [110]. This precession frequency is named as "Lamar frequency" and depends upon two factors, one is gyromagnetic ratio (γ) while other is applied field $B_\circ$, and represented as

$$w_\circ = \gamma B_\circ$$

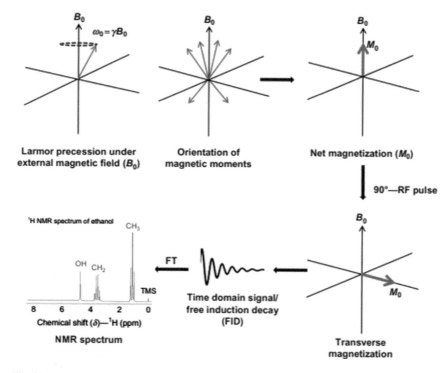

Fig. 20 Principle of NMR spectroscopy [110]

When radio frequency (MHz) radiations are fallen on the sample, they cause the nucleus to absorb specific energy; as a result, the angle of precession changes. The absorption of energy takes place only when the incident frequency matches with the precession's frequency. When an external applied field is removed, as a result, the nucleus returns back to its lower energy state and releases extra energy. Such disruption of energy is studied and recorded to obtain fingerprint characterization of a given NP's molecule [110]. On the basis of this principle, different experiments of NMR spectroscopy such as 1D, 2D and multidimensional are developed to study different properties including structure, confinement, molecule's dynamics, etc. NMR is mostly implemented on the liquid state solutions, but due to advancement in technologies and experimental techniques including high resolution (HR–MAS), cross-polarization magic angle spin (CP–MAS), etc., NMR spectroscopy is also implemented on solid state [111–116]. Some of the NMR includes heteronuclear single-quantum correlation, exchange spectroscopy, magnetic resonance scanning, diffusion spectroscopy, correlation spectroscopy, etc. NMR is considered as the most suitable direct surface area measurement technique due to its property of characterization of NMs in solution [117].

3.8.2   NMR Relaxation Time

Rather than examining a spectrum, a substitute methodology is to gauge relaxation times (RT) from pulses of NMR. There are two RT; one is longitudinal magnetic relaxation time (LMRT), i.e., represented as T1, while the other one is transverse magnetic relaxation time (TMRT), i.e., represented as T2.

LMRT (T1) belongs to spin–lattice relaxation at which proton energy is transferred to higher energy state. In solids, this occurs at faster rate as neighboring atoms are able to accept energy of 1 ms order, while in case of liquids, its order may be of 10 s. TMRT (T2) belongs to spin relations where a photon's energy is transferred to higher energy state's spin-off to its neighboring proton. There are a variety of relaxation sub-categories depending upon pulse sequences including decay measurement of TMRT which are done via Hahn-echo sequence (T2), and this transverse magnetization decay is recorded by utilization of solid-echo decay (T2e). Due to varying environment, the substance relaxation times (T1&T2) are affected. Short and long relaxation times indicate hard or more solid like environments and soft or plasticized environments, respectively. Hence, materials including 1D, 2D or multiphase NMs with different molecular environments are characterized by utilizing relaxation time. The RT indicated by the exponential distribution as well as multiple relaxation times indicate superposition of distribution. The relaxation distribution amplitude's sum is given as an individual relaxation distribution's function of Laplace transformation. Individual relaxation function can be demagnetized in an inverse Laplace transformation. Relaxation times (T1&T2) give us information about characterization within a substance [55]. The magnetic properties of ^1H using NRM RT of iron oxide NPs indicate that it is suitable for utilization in MRI as a contrast.

4   Conclusion

The chapter is divided into three sections; the first part discussed the introduction of nanostructures and their classification. The second part dealt with fundamental properties possessed by nanostructures. In physical properties of nanostructures, thermal properties of nanostructures suggested that the morphology of nanostructures can be controlled through annealing and transport of electrons. Optical properties such as band structure and transitions such as emission and absorption are dependent on the quantum confinement and surface effects. Electronic properties suggest that the density of device increases by decreasing its size. Mechanical properties such as hardness, fatigue strength, fatigue toughness and bending strength are very important for practical implementation of nanostructures. Transport properties depend on the size and length of the system. Surface area and surface energy of the system are related to the grain size of nanostructures, while chemical potential depends on the radius of curvature. Reduction of surface energy due to large surface area is very critical and can be achieved through surface restructuration, surface adsorption and surface relaxation. Variations in size of nanostructures have revolutionized the thermodynamic, electronic, spectroscopic, structural and chemical behavior of these systems. In addition, morphology has exceptional effect upon chemical and physical properties of nanostructures. Shape of the nanostructure is also correlated with their properties and molecular interaction in contrast with the bulk materials.

Different characterization techniques help in exploring structure, morphology and other properties of nanostructured materials. Photon-optical spectroscopy includes Raman and fluorescence spectroscopies with resolution limit up to 0.25 μm, utilizing confocal probe version that mostly characterize carbon-based nanostructures. Nano-powdered oxides and multilayered nanostructures are analyzed by Raman spectroscopy, while SWCNTs are analyzed by fluorescent spectroscopy. Electro-optical imaging of nanostructures include TEM, SEM and STEM. TEM provides topographical visualization of nanostructures. Internal structure of specimens is imaged by TEM, while SEM forms the images of upper or near surface, and bulk specimen's topography. TEM cannot create contrast in backscattering mode, but this backscattering image is very popular in SEM. A field emission cathode is implemented in SEM to improve spatial resolution which is FESEM. Scanning probe microscopy (SPM) is a surface characterization technique through which topography of nanostructures can be explored. The main members of the SPM family are scanning tunneling microscopy (STM) and atomic force microscopy (AFM). STM forms images of the surface with greater atomic accuracy with a tip. High-resolution images of the conductive or semiconducting surfaces are formed by STM, with a lateral resolution of 0.1–0.01 nm in depth. STM forms images of organic molecules like tungsten trioxide (WO_3) thin films, when placed upon conducting substrates in a vacuum environment. Atomic force microscopy has the advantage that different surfaces or environmental factors do not limit its good performance. The AFM is used to get information related to different parameters including roughness, surface texturization, morphology, friction, magnetic, elastic properties, etc. Moreover, AFM is cost effective and has resolution of about "0.2 nm." Molecular modeling is a theoretical computation-based model in which simulation of different NMs is done in 3D to deliberate different properties like bulk, electronic, optical, spectroscopic, geometries, etc. XRD is used to give significant information related to physicochemical properties of NMs. Small-angle X-ray scattering (SAXS) is observed at very small angles typically $0.1°–10°$. SAXS is limited to crystalline materials such as monocrystalline and polycrystalline. WAXS is utilized to measure crystalline size of nanoparticles. Different NMs including microsphere NiO and $NiCO_2O_4$ nanowires are characterized by WAXS. X-ray topography is used for detection of defects in NMs structures or the localization of corrosion. For XPS, ultra-high voltage is required. Different NMs including $Au@SiO_2$ NPs, $Au@Ag$ core–shell NPsCdS$@Ag_2S$ core/shell NPs, etc., are characterized by XPS.

References

1. Jeevanandam, J., Ahmad, B., Chan, Y.S, Alain, D., Michael, K.D.: Review on nanoparticles and nanostructured materials: history, sources, toxicity and regulations. Beilstein J. Nanotechnol. **9**(1):1050–1074
2. Gopalakrishnan, S., Narendar, S.: Wave Propagation in Nanostructures: Nonlocal Continuum Mechanics Formulations. Springer Science & Business Media (2013)
3. Toumey, C.: Plenty of room, plenty of history. Nat. Nanotechnol. **4**(12), 783–784 (2009)
4. Shanks, R.A.: Characterization of nanostructured materials. Nanostructured Polymer Blends **1**(1), 15–31 (2014)
5. Gleiter, H.: Nanostructured materials: basic concepts and microstructure. Acta Materialia **48**(1), 1–29
6. O'connell, M.J.: Carbon Nanotubes: Properties and Applications. CRC Press, Tayler and Francis, Boca Raton (2006)
7. Koch, C.C.: Nanostructured Materials: Processing, Properties and Applications. William Andrew (2006)
8. Gusev, A.I., Rempel', A.A.: Nanocrystalline materials. Cambridge International Science Publishing, Cambridge (2004)
9. Logothetidis, S.: Nanostructured Materials and Their Applications. Springer Science & Business Media, Berlin (2012)
10. Cao, H., Synthesis and Applications of inorganic nanostructures. Wiley, New York (2017)
11. Wang, Z.M.: One-Dimensional Nanostructures, vol. 3. Springer Science & Business Media, Berlin (2008)
12. Chen, S., Li, W., Li, X., Yang, W.: One-dimensional SiC nanostructures: designed growth, properties, and applications. Prog. Mater Sci. **104**, 138–214 (2019)
13. Boddula, R., Asiri, A.M.: Self-standing Substrates: Materials and Applications. Springer Nature, Berlin (2019)
14. Cheng, X.: Nanostructures: fabrication and applications. Nanolithography 348–375
15. Nayak, B.B., Behera, D., Mishra, B.K.: Synthesis of silicon carbide dendrite by the arc plasma process and observation of nanorod bundles in the dendrite arm. J. Am. Ceram. Soc. **93**(10), 3080–3083 (2010)
16. Mann, A.K., Skrabalak, S.E.: Synthesis of single-crystalline nanoplates by spray pyrolysis: a metathesis route to Bi_2WO_6. Chem. Mater. **23**(4), 1017–1022 (2011)
17. Siril, P.F., Ramos, L., Beaunier, P., Archirel, P., Etcheberry, A., Remita, H.: Synthesis of ultrathin hexagonal palladium nanosheets. Chem. Mater. **21**(21), 5170–5175 (2009)
18. Vizireanu, S., Stoica, S.D., Luculescu, C., Nistor, L.C., Mitu, B., Dinescu, G.: Plasma techniques for nanostructured carbon materials synthesis. A case study: carbon nanowall growth by low pressure expanding RF plasma. Plasma Sources Sci. Technol. **19**(3), 034016 (2010)
19. Jung, S.H., Oh, E., Lee, K.H., Yang, Y., Park, C.G., Park, W., Jeong, S.H.: Sonochemical preparation of shape-selective ZnO nanostructures. Cryst Growth Design **8**(1), 265–269 (2008)
20. Thangadurai, T.D., et al.: Fundamentals of nanostructures. In: Nanostructured Materials, pp. 29–45. Springer, Cham (2020)
21. Wang, H., Li, M., Li, X., Xie, K., Liao, L.: Preparation and thermal stability of nickel nanowires via self-assembly process under magnetic field. Bull. Mater. Sci. **38**(5), 1285–1289 (2015)
22. Bhushan, B.: Nanotribology and nanomechanics. Wear **259**(7–12), 1507–1531 (2005)
23. Schaefer, H.E.: Nanoscience: The Science of the Small in Physics, Engineering, Chemistry, Biology and Medicine. Springer Science & Business Media, Berlin (2010)
24. Singh, M.R., Cox, J.D., Brzozowski, M.: Photoluminescence and spontaneous emission enhancement in metamaterial nanostructures. J. Phys. D Appl. Phys. **47**(8), 085101 (2014)
25. Spanier, J.E.: One-dimensional semiconductor and oxide nanostructures. Nanotubes Nanofibers 199–232 (2006)
26. Schroer, M.D., Petta, J.R.: Correlating the nanostructure and electronic properties of InAs nanowires. Nano Lett. **10**(5), 1618–1622 (2010)

27. Joyce, H.J., Boland, J.L., Davies, C.L., Baig, S.A., Johnston, M.B.: A review of the electrical properties of semiconductor nanowires: insights gained from terahertz conductivity spectroscopy. Semicond. Sci. Technol. **31**(10), 103003 (2016)
28. Bhushan, B.: Mechanical properties of nanostructures. In: Nanotribology and Nanomechanics I, p. 527-584. Springer, Berlin (2011)
29. Thangadurai, T.D., Maniubaashini, N., Thomas, S., Maria, H.J.: Properties of nanostructured materials. In: Nanostructured Materials, pp. 77–95. Springer, Chem (2020)
30. Sabirov, I., Enikeev, N.A., Murashkin, M.Y., Valiev, R.Z.: Bulk Nanostructured Materials with Multifunctional Properties, vol. 118. Springer International Publishing, Berlin (2015)
31. Fougere, G., Riester, L., Ferber, M., Weertman, J.R., Sieqel, R.W.: Young's modulus of nanocrystalline Fe measured by nanoindentation. Mater. Sci. Eng., a **204**(1–2), 1–6 (1995)
32. Andrievski, R.A.: Nanocrystalline high melting point compound-based materials. J. Mater. Sci. **29**(3), 614–631 (1994)
33. Bhushan, B.: Handbook of Micro/Nanotribology. CRC Press, Boca Raton (1998)
34. Namazu, T., Isono, Y., Tanaka, T.: Evaluation of size effect on mechanical properties of single crystal silicon by nanoscale bending test using AFM. J. Microelectromech. Syst. **9**(4), 450–459 (2000)
35. Lawn, B.R., Evans, A.G., Marshall, D.B.: Elastic/plastic indentation damage in ceramics: the median/radial crack system. J. Am. Ceram. Soc. **63**(9–10), 574–581 (1980)
36. Pippan, R., Hohenwarter, A.: The importance of fracture toughness in ultrafine and nanocrystalline bulk materials. Mater. Res. Lett. **4**(3), 127–136 (2016)
37. Sundararajan, S., Bhushan, B.: Development of AFM-based techniques to measure mechanical properties of nanoscale structures. Sens. Actuators A **101**(3), 338–351 (2002)
38. Cao, G.: Nanostructures and nanomaterials: synthesis, properties and applications. World Sci. (2004)
39. Goldstein, A.N., Echer, C.M., Alivisatos, A.P.: Melting in semiconductor nanocrystals. Science **256**(5062), 1425–1427 (1992)
40. VanHove, M.A., Weinberg, W.H., Chan, C.M.: Low-Energy Electron Diffraction: Experiment, Theory and Surface Structure Determination, vol. 6. Springer Science & Business Media, Berlin (2012)
41. Vook, R.W.: Structure and growth of thin films. Int. Metals Rev. **27**(1), 209–245 (1982)
42. Iler, K.R.: The chemistry of silica. In: Solubility, Polymerization, Colloid and Surface Properties and Biochemistry of Silica (1979)
43. La Mer, V.K., Gruen, R.: A direct test of Kelvin's equation connecting vapour pressure and radius of curvature. Trans. Faraday Soc. **48**, 410–415 (1952)
44. Schöll, E.: Theory of Transport Properties of Semiconductor Nanostructures, vol. 4. Springer Science & Business Media (2013)
45. Siegel, R.W., Hu, E.: Nanostructure Science and Technology: R&D Status and Trends in Nanoparticles, Nanostructured Materials and Nanodevices. Springer Science & Business Media, Berlin (1999)
46. Thangadurai, T.D., Maniubaashini, N., Thomas, S., Maria, H.J.: Physics and chemistry of nanostructures. In: Nanostructured Materials, pp. 47–53. Springer, Cham (2020)
47. Gobre, V.V., Tkatchenko, A.: Scaling laws for van der Waals interactions in nanostructured materials. Nat. Commun. **4**(1), 1–6 (2013)
48. Thangadurai, T.D., Maniubaashini, N., Thomas, S., Maria, H.J.: Characterization and technical analysis of nanostructured materials. In: Nanostructured Materials, pp. 119–128. Springer, Cham (2020)
49. Myhra, S., Rivière, J.C.: Characterization of Nanostructures. CRC Press, Boca Raton (2012)
50. Otto, A.: What is observed in single molecule SERS, and why? J. Raman Spectrosc. **33**(8), 593–598 (2002)
51. Nichols, M.E., Seubert, C.M., Weber, W.H., Gerlock, J.L.: A simple Raman technique to measure the degree of cure in UV curable coatings. Prog. Org. Coat. **43**(4), 226–232 (2001)
52. Kumar, P.S., Pavithra, K.G., Naushad, M.: Characterization techniques for nanomaterials. In: Nanomaterials for Solar Cell Applications, pp. 97–124. Elsevier, Amsterdam (2019)

53. Sharma, A., Schulman, S.G.: Introduction to Fluorescence Spectroscopy, vol. 13. Wiley, New York (1999)
54. Weisman, R.B.: Fluorimetric characterization of single-walled carbon nanotubes. Anal. Bioanal. Chem. **396**(3), 1015–1023 (2010)
55. Pennycook, S.J.: Z-contrast STEM for materials science. Ultramicroscopy **30**(1–2), 58–69 (1989)
56. Choudhary, O.P., Kalita, P.C., Doley, P.C., Kalita, A.: Scanning electron microscope: advantages and disadvantages in imaging components. Life Sci. Leaflets **85**, 1877–1882 (2017)
57. Marturi, N.: Vision and visual servoing for nanomanipulation and nanocharacterization in scanning electron microscope. In: CCSD (2013)
58. Fritzinger, B., Moreels, I., Lommens, P., Koole, R., Hens, Z., Martins, J.C.: In situ observation of rapid ligand exchange in colloidal nanocrystal suspensions using transfer NOE nuclear magnetic resonance spectroscopy. J. Am. Chem. Soc. **131**(8), 3024–3032 (2009)
59. Brabazon, D., Raffer, A.: Advanced characterization techniques for nanostructures. In: Emerging Nanotechnologies for Manufacturing, pp. 59–91. Elsevier, Amsterdam (2010)
60. Liu, J.: Advanced electron microscopy characterization of nanostructured heterogeneous catalysts. Microsc. Microanal. **10**(1), 55–76 (2004)
61. Maffeis, T.G.G., Yung, D., LePennec, L., Penny, M.W., Coblev, R.J., Comimi, E., Shervealieri, G., Wilks, S.P.: STM and XPS characterisation of vacuum annealed nanocrystalline WO3 films. Surf. Sci. **601**(21), 4953–4957 (2007)
62. Barceló, D., Farré, M.: Analysis and Risk of Nanomaterials in Environmental and Food Samples, vol. 59. Newnes (2012)
63. Pantano, C.G.., Bojan, V.J., Smay, G.: AFM analysis of hot-end coatings on glass containers. Glass Res. **9**(2), 12–13 (1999)
64. Giessibl, F.J.: Advances in atomic force microscopy. Rev. Mod. Phys. **75**(3), 949 (2003)
65. Neubauer, E., Eisenmenger-Sittner, C., Bangert, H., Korb, G., Thomastik, C.: AFM and Auger investigations of as-deposited and heat treated copper coatings on glassy carbon surfaces with chromium and molybdenum intermediate layers. Surf. Coat. Technol. **180**, 496–499 (2004)
66. Neubauer, E., Neubauer, E., Eisenmenger-Sittner, C., Bangert, H., Korb, G.: AFM and AUGER investigations of as-deposited and heat treated copper coatings on glassy carbon surfaces with titanium intermediate layers. Vacuum **71**(1–2), 293–298 (2003)
67. Aliofkhazraei, M., Ali, N.: AFM Applications in Micro/Nanostructured Coatings. Elsevier, Amsterdam (2014)
68. Ray, S.S.: Clay-containing polymer nanocomposites: from fundamentals to real applications. Newnes (2013)
69. Sitterberg, J., Oxcetin, A., Ehrhardt, C., Bakowsky, U.: Utilising atomic force microscopy for the characterisation of nanoscale drug delivery systems. Eur. J. Pharm. Biopharm. **74**(1), 2–13 (2010)
70. Ray, S.S.: Environmentally Friendly Polymer Nanocomposites: Types, Processing and Properties. Elsevier, Amsterdam (2013)
71. Rane, S., Beaucage, G., Satkowski, M.M.: Morphological study of polyhydroxyalkanoates and their blends. Polymeric Mater. Sci. Eng. (USA) **80**, 402–403
72. Barbosa, N.S.V., de Almeida Lima, E.R., Tavares, F.W.: Molecular Modeling in Chemical Engineering. Elsevier, Amsterdam (2017)
73. Andrew, R.L.: Molecular Modeling Principles and Applications, 2nd ed. Pearson Education Upper Saddle River (2001)
74. Leach, A.R., Leach, A.R.: Molecular Modelling: Principles and Applications. Pearson Education, Upper Saddle River (2001)
75. Gu, Y., Li, M.: Molecular modeling. In: Handbook of Benzoxazine Resins, pp. 103–110. Elsevier, Amsterdam (2011)
76. Kohn, W., Sham, L.J.: Self-consistent equations including exchange and correlation effects. Phys. Rev. **140**(4A), A1133 (1965)

77. Fernandes, M.M., Baeyens, B., Beaucaire, C.: Radionuclide Retention at Mineral–Water Interfaces in the Natural Environment. Radionuclide Behaviour in the Natural Environment, pp. 261–301. Elsevier, Amsterdam (2012)
78. Cejka, J., van Bekkum, H.: Zeolites and ordered mesoporous materials: progress and prospects: the 1st FEZA School on Zeolites, Prague, Czech Republic, August 20–21, vol. 157. Gulf Professional Publishing (2005)
79. Mourdikoudis, S., Pallares, R.M., Thanh, N.T.: Characterization techniques for nanoparticles: comparison and complementarity upon studying nanoparticle properties. Nanoscale **10**(27), 12871–12934 (2018)
80. Giannini, C., Siliqi, D., Altamura, D.: Nanomaterial characterization by X-ray scattering techniques. In: Nanocomposites: In Situ Synthesis of Polymer-Embedded Nanostructures, pp. 209–222 (2013)
81. Kumar, C.S.: Surface Science Tools for Nanomaterials Characterization. Springer, Berlin (2015)
82. Bindu, P., Thomas, S.: Estimation of lattice strain in ZnO nanoparticles: X-ray peak profile analysis. J. Theor. Appl. Phys. **8**(4), 123–134 (2014)
83. Peck, M.A., Langell, M.A.: Comparison of nanoscaled and bulk NiO structural and environmental characteristics by XRD, XAFS, and XPS. Chem. Mater. **24**(23), 4483–4490 (2012)
84. Harrison, P.G., Willett, M.J.: Tin oxide surfaces. Part 19.—Electron microscopy, X-ray diffraction, auger electron and electrical conductance studies of tin (IV) oxide gel. J. Chem. Soc. Faraday Trans. Phys. Chem Condensed Phases **85**(8), 1907–1919
85. Wang, J., Qin, S.: Study on the thermal and mechanical properties of epoxy–nanoclay composites: the effect of ultrasonic stirring time. Mater. Lett. **61**(19–20), 4222–4224 (2007)
86. Mi, Y., Chen, X., Guo, Q.: Bamboo fiber-reinforced polypropylene composites: crystallization and interfacial morphology. J. Appl. Polym. Sci. **64**(7), 1267–1273 (1997)
87. Bishnoi, A., Kumar, S., Joshi, N.: Wide-angle X-ray diffraction (WXRD): technique for characterization of nanomaterials and polymer nanocomposites. In: Microscopy Methods in Nanomaterials Characterization, pp. 313–337. Elsevier, Amsterdam (2011)
88. Srivastava, O.N., Srivastava, A., Dash, D., Singh, D.P., Yadav, R.M., Mishra, P.R., Singh, J.: Synthesis, characterizations and applications of some nanomaterials (TiO$_2$ and SiC nanostructured films, organized CNT structures, ZnO structures and CNT-blood platelet clusters). Pramana **65**(4), 581–592 (2005)
89. Singh, M., Sinha, I,, Singh, A.K., Mandal, R.K. LSPR and SAXS studies of starch stabilized Ag–Cu alloy nanoparticles. Colloids Surf. A Physicochem. Eng. Aspects. **384**(1–3), 668–674 (2011)
90. Bulavin, L., Kutsevol, N., Chumachenko, V., Soloviov, D., Kuklin, A, Marvnin, A.: SAXS combined with UV-vis spectroscopy and QELS: accurate characterization of silver sols synthesized in polymer matrices. Nanoscale Res. Lett. **11**(1), 35 (2016)
91. Zhao, Y., Saiio, K., Takenake, M., Koizumi, S., Hashimoto, T.: Time-resolved SAXS studies of self-assembling process of palladium nanoparticles in templates of polystyrene-block-polyisoprene melt: effects of reaction fields on the self-assembly. Polymer **50**(12), 2696–2705 (2009)
92. Li, T., Senesi, A.J., Lee, B.: Small angle X-ray scattering for nanoparticle research. Chem. Rev. **116**(18), 11128–11180
93. Chaurand, P., Liu, W., Borschneck, D., Levard, C., Auffan, M., Paul, E., Collin, B., Kieffer, I., Lanone, S., Rose, J., Perrin, J.: Multi-scale X-ray computed tomography to detect and localize metal-based nanomaterials in lung tissues of in vivo exposed mice. Sci. Rep. **8**(1), 1–11 (2018)
94. Withers, P.J.: X-ray nanotomography. Mater. Today **10**(12), 26–34 (2007)
95. Lozano-Perez, S.: Characterization Techniques for Assessing Irradiated and Ageing Materials in Nuclear Power Plant Systems, Structures and Components (SSC). Understanding and Mitigating Ageing in Nuclear Power Plants, pp. 389–416. Elsevier, Amsterdam (2010)

97. Poulsen, H.F.: Three-Dimensional X-ray Diffraction Microscopy: Mapping Polycrystals and Their Dynamics, vol. 205. Springer Science & Business Media, Berlin (2004)

98. Matthew, J.: Surface analysis by Auger and X-ray photoelectron spectroscopy. In: D. Briggs and JT Grant (eds). IMPublications, Chichester, UK and SurfaceSpectra, Manchester, UK, 900pp (2003). ISBN 1-901019-04-7. Surf Interface Anal. Int. J. **36**(13), 1647–1647 (2004). Devoted to the development and application of techniques for the analysis of surfaces, interfaces and thin films

99. King, A., Johnson, G., Engerlberg, D., Ludwing, W., Marrow, J.: Observations of intergranular stress corrosion cracking in a grain-mapped polycrystal. Science **321**(5887), 382–385 (2008)

100. Sarma, D.D., Santra, P.K., Mukherjee, S., Naq, A.: X-ray photoelectron spectroscopy: a unique tool to determine the internal heterostructure of nanoparticles. Chem. Mater. **25**(8), 1222–1232 (2013)

101. Hota, G., Idage, S., Khilar, K.C.: Characterization of nano-sized CdS–Ag2S core-shell nanoparticles using XPS technique. Colloids Surf. A Physicochem. Eng. Aspects **293**(1–3), 5–12 (2007)

102. Wang, Y.C., Engelhard, M.H., Baer, D.R., Castner, D.G.: Quantifying the impact of nanoparticle coatings and nonuniformities on xps analysis: gold/silver core–shell nanoparticles. Anal. Chem. **88**(7), 3917–3925 (2016)

103. Tunc, I., Demirok, U.K., Suzer, S., Correa-Duatre, M.A., Liz-Marzan, L.M.: Charging/discharging of Au (core)/silica (shell) nanoparticles as revealed by XPS. J. Phys. Chem. B **109**(50), 24182–24184 (2005)

104. Zeng, D.W., Xie, C.S., Zhu, B.L., Jianq, R., Chen, X., Song, W.L., Wang, J.B., Shi, J.: Controlled growth of ZnO nanomaterials via doping Sb. J. Cryst. Growth **266**(4), 511–518 (2004)

105. Rao, C., Biswas, K.: Characterization of nanomaterials by physical methods. Annu. Rev. Anal. Chem. **2**, 435–462 (2009)

106. Khan, S.H.: Green nanotechnology for the environment and sustainable development. In: Green Materials for Wastewater Treatment, pp. 13–46. Springer, Berlin

107. Baer, D.R., Engelhard, M.H., Johson, G.E., Laskin, J., Mueller, K., Munusamy, P., Thevuthasan, S., Wang, H., Washton, N., Elder, A.: Surface characterization of nanomaterials and nanoparticles: important needs and challenging opportunities. J. Vac. Sci. Technol. **31**(5), 050820 (2013)

108. Marbella, L.E., Millstone, J.E.: NMR techniques for noble metal nanoparticles. Chem. Mater. **27**(8), 2721–2739 (2015)

109. Roming, M., Feldmann, C., Avadhut, Y.S., der Gunne, J.S.A.: Characterization of noncrystalline nanomaterials: NMR of zinc phosphate as a case study. Chem. Mater. **20**(18), 5787–5795 (2008)

110. Agarwal, N., Nair, M.S., Mazumder, A. Poluri, K.M.: Characterization of nanomaterials using nuclear magnetic resonance spectroscopy. In: Characterization of Nanomaterials, pp. 61–102. Elsevier, Ansterdam

111. Zhang, J., Higashi, K., Limwikrant, W., Moribe, K., Yamamoto, K.: Molecular-level characterization of probucol nanocrystal in water by in situ solid-state NMR spectroscopy. Int. J. Pharm. **423**(2), 571–576 (2012)

112. Lo, A.Y., Sudarsan, V., Sivakumar, S., van Veggel, F., Schurko, R.W.: Multinuclear solid-state NMR spectroscopy of doped lanthanum fluoride nanoparticles. J. Am. Chem. Soc. **129**(15), 4687–4700 (2007)

113. Odin, C., Ameline, J.: Orientational deconvolution of two-dimensional static disorder by a Tikhonov regularized method for 2H solid state NMR of nano-tubular-oriented structures. Solid State Nucl. Magn. Reson. **27**(4), 257–265 (2005)

114. Mehnert, W., Mäder, K.: Solid lipid nanoparticles: production, characterization and applications. Adv. Drug Deliv. Rev. **64**:83–101

115. Jenning, V., Thünemann, A.F., Gohla, S.H.: Characterisation of a novel solid lipid nanoparticle carrier system based on binary mixtures of liquid and solid lipids. Int. J. Pharm. **199**(2), 167–177 (2000)

116. Jenning, V., Mäder, K., Gohla, S.H.: Solid lipid nanoparticles (SLN^{TM}) based on binary mixtures of liquid and solid lipids: a 1H-NMR study. Int. J. Pharm. **205**(1–2), 15–21 (2000)
117. Baalousha, M., Lead, J.: Characterization of Nanomaterials in Complex Environmental and Biological Media. Elsevier, Amsterdam (2015)

Nanostructures: A Solution to Quantum Computation and Energy Problems

Asma Ayub, Aleena Shoukat, Muhammad Bilal Tahir, Hajra Kanwal, Muhammad Sagir, and Muhammad Rafique

Abstract Nanostructures are one of the fundamental aspects of nanotechnology. Various types of nanostructures play a significant role in quantum informatics and quantum computation. Quantum computation involves the storage and manipulation of information by means of quantum bits. For this purpose, several nanostructures having a short dechoerence time are being fabricated such as graphene quantum dots and Josephson junctions. This chapter deals with the role of nanostructures in quantum informatics and in energy sources. It is anticipated that energy transportation and energy storage capacity can be improved by utilizing nanomaterials.

Keywords Quantum computation · Quantum bits · Josephson junctions · Energy solution · Semiconductor nanostructures · Quantum dots · Energy resources

1 Nanostructures

Nanostructures oblige a great role in a lot of industries and fields due to their surprisingly small size and high surface area [1]. Nanostrutures are being fabricated since past few decades because of having novel properties as optical, structural, morphological, mechanical, etc. [2]. The properties of solid state particles, such as electrical, thermal, and optical, can be controlled by proper quantum confinement of electrons. Nanostructures play a protuberant role in informatics, where a

A. Ayub · A. Shoukat · H. Kanwal
Department of Physics, Faculty of Sciences, University of Gujrat, Gujrat 50700, Pakistan

M. B. Tahir (✉)
Department of Physics, Khwaja Fareed University of Engineering and Information Technology, Rahim Yar Khan, Pakistan
e-mail: m.bilaltahir@kfueit.edu.pk

M. Sagir
Department of Chemical Engineering, Khwaja Fareed University of Engineering and Information Technology, Rahim Yar Khan, Pakistan

M. Rafique
Department of Physics, University of Sahiwal, Sahiwal, Pakistan

© The Author(s), under exclusive license to Springer Nature Singapore Pte Ltd. 2021
M. B. Tahir et al. (eds.), *Nanotechnology*,
https://doi.org/10.1007/978-981-15-9437-3_4

lot of efforts have been made to achieve nano-sized storage units. Quantum wells known as two-dimensional nanostructures have been attained much attention by researchers because of their applications in the field of quantum informatics [3, 4]. Zero-dimensional and one-dimensional nanostructures are also studied extensively in the field of nanotechnology [5, 6].

The advancement and innovations in the field of nanotechnology are being made since fourth century AD by Romans [7].

1.1 Semiconductor Nanostructures

Several techniques are utilized for the development of semiconductor-based low-dimensional systems. MBE known as molecular beam epitaxy is extensively used in semiconductors devices [8]. Utilization of advanced techniques such as MBE and MOCVD allows one to develop low-dimensional semiconductor devices with the precision of single atomic layer. These low-dimensional semiconductor devices are dependent on their intrinsic properties and are utilized in optic-electronics [9].

Hetro-structures are spatially varying semiconductor nanostructures. The variation in hetro-structures composition in one dimension leads to the two-dimensional homogenous semiconductor layers. Quantum wells are the type of hetro-structures in which a low band gap semiconductor is sandwiched between two layers of larger band gap semiconductors, thus forming a hetro-junction.

Quantum wires are another type of semiconductor nanostructure produced via micro-structuring of quantum wells. Quantum dots play a main role in quantum computation, quantum information processing, and quantum informatics. They can be produced via three-dimensional confinement of carriers in quantum wells [10, 11].

2 Quantum Informatics

With the advancement in the expertise of quantum information processing, the realization of quantum computing devices has been made possible. Quantum computation is the new face of computation and has more potential as compared to other computing techniques [12].

Any computation requires bits just as binary digits are used in regular computation. Quantum computation is realized through proper quantum bits [13].

Semiconductor quantum dots are one of the examples of quantum bits that are frequently used in the quantum informatics. Electrons that reside in the quantum dots are encrypted with information. This encryption is based on the spin and orbital degree of freedom of excitons [14].

Quantum dots play a pivotal role in a quantum framework such as they are used in THz lasers. Quantum informatics have revolutionized the optical system.

Much progress has been made on the fabrication of new techniques and strategies to incorporate quantum dots in optical systems [15].

Quantum dots are made to interact through their phonon mode. The interaction is characterized by the size, type, and photonic structure of the quantum dots involved. The chapter deliberates the use of quantum dots in nanotechnology via nanostructures [12].

2.1 The Role of Nanostructures in Quantum Informatics

This section discusses the role of different nanostructures in the various fields of quantum informatics such as quantum computation.

2.1.1 Quantum Computation Through Semiconductor Nanostructures

Quantum informatics have garnered various attentions in many fields ranging from optics to condensed matter physics via quantum computation and quantum information processing [16, 17]. Quantum computation is the hot topic in the research fields. Companies are competing to build the quantum computer, in order to attain an algorithm for resolving extremely large factors that are not solvable by classical algorithm [18].

Another goal of quantum computation is to increase the speed of algorithm that can be made possible through the ideas of quantum mechanics such as superposition principle. It is predicted that quantum computer can perform computation with quantum bits (q-bits) and can factorize much faster than its classical counterpart. Shor's factorizing algorithm that led to the development of quantum computation states that an error correction can be realized in a quantum system [19].

Q-bits play a significant role in quantum computation. Many propositions of using electrons or the spin of semiconductor as q-bits have been made [20]. Quantum dots spin [12, 21] and semiconductors nuclear spin [22, 23] have also been proposed as a possible contender for q-bits.

The donor electrons in n-type semiconductor can also be considered as a potential q-bit. The wave function of these donor electrons is similar to semiconductors. Hence, they can be used in a quantum computer as a qubit.

Microelectronic devices play an important role in quantum computation since there exists a connection between quantum devices and microelectronic devices. Doping has played a significant role in the industry of microelectronics over ther past 50 years. Due to miniaturization of integrated circuits and transistors, the sensitivity and configuration of doping impurities have been increased [24].

Due to donor electrons, Kane, in 1998, proposed a donor semiconductor-based quantum computer [20]. In this semiconductor-based quantum computer, the monovalent impurities act as quantum bits. Both orbital and spin degree of excitons can be utilized as quantum bits in nanostructures. Semiconductor quantum dots having radius of about 10 nm are used as orbital q-bit with their quantized orbital and spin degree of freedom. These quantum dots are useful because of their specific measurements such as charged states that are measured through SET experimentally [25]. The downsides of such quantum bit are its small dechoerence time for the quantum error corrections and its susceptibility to environment.

The coupling of two q-bit gate operations is a requirement for quantum computation. Charged-bit coupling involves large range of dipolar coupling that is similar to the inter q-bit coupling [26, 27]. Due to this similarity, the increment in dechoerence is difficult because of scaling up. The scaling up of quantum computer then leads to large q-bit dipolar coupling.

The traditional solid state quantum computation has the advantage of scalability, but quantum dots-based quantum computation is still attractive due to its entanglement and dechoerence properties [28].

Spin q-bits as compared to orbital charged q-bits have advantage of having large spin coherence time at low temperature but a disadvantage such as measurement of spin of a single electron in semiconductor nanostructures. The large coherence time of electron spin q-bits is the reason of their application in quantum computation as quantum dots and donor electrons. Spin q-bits have an exchange gate, with short range, that is responsible for inter q-bit coupling. The short range enables the exchange gates to control over the two q-bit gates. The quantum computer based on such q-bits does not have the problem of scaling up, since coupling of the q-bits is independent of the number of q-bits present.

Spectral diffusion is the dechoerence mechanism that is employed in semiconductor nanostructures for electron spin. The nuclear spin scale is very small and is hard to extract from a sample. Semiconductor nanostructures have larger coherence time as compared to gallium arsenide nanostructures [29].

The use of semiconductor nanostructures in the versatile field of quantum computation is an indication of the rapid progress in nanotechnology and the computation world.

2.1.2 Quantum Information Processing with Nanostructures

In the recent past few years, a lot of research has been done in the field of quantum informatics that suggests the quantum bits as an important factor for quantum computation and information processing [30]. For practical applications in quantum information, processing the isolation of quantum bits is significant for controlled measurement and precision [31].

The proper tuning of the quantum bits interaction is very crucial. Involved quantum operations should be observable for any gate operations. The scalability of quantum system is a very important factor in quantum information processing.

The quantum information processing is a phenomenon that can be realized with particles such as atoms, ions, and photons [32].

The ideal setup for quantum computation and communication consists of well-defined coupling (strong or weak) between atoms or ions and photons. Trapped ions can also be employed in information processing as well as in solid state systems that are used for large-scale processing. Multiple solid state systems are proposed for application in quantum information processing [33–39]. These solid state systems require complex quantum system for processing with similar short dechoerence and size problem. The information processing at a time less than dechoerence time is possible in these solid state systems called nanostructures.

Nanostructures either artificial or natural and quantum dots are an active candidate for quantum information processing because of the advancement in their nanofabrication. Semiconductors quantum dots are nanostructures that have a size in the range of nanometers [40, 41]. The optical and electrical properties of such nanostructures or quantum dots can be controlled via quantum confinement because of their extremely small size. Quantum dots behave like individual atoms but have a difference in their sizes and energies. Quantum dots have smaller size and larger energy as compared to individual atoms [40].

Quantum dots are employed in quantum informatics because of their much smaller size than atoms and localization. Quantum dots can store quantum information via two ways as in the form of electron spin q-bits or charged excitons bits. Coupling of two quantum dots results in a new molecule that can also serve as a q-bit and can be used for storing quantum information.

Quantum information processing based on quantum dots was explained via the interactions of spin in the past, and such interactions were controlled using electrical gates [35, 42]. It is now considered that quantum system with optically controlled interactions is more reliable because laser technology allows the quantum communication in the desired coherence time. The interaction of photon and quantum dot can simplify the quantum communication over larger distances [43]. Quantum dots can be integrated over nano-cavities for quantum electrodynamics applications [21, 44]. The existence of ultrafast laser with very large pulse energies, widths, repetition rates, and wavelengths has made possible to control the coherence in semiconductor nanostructures before the transient regime [43]. For the implementation of quantum communication and computation, semiconductor nanostructures that are equipped with ultrafast optical technology can serve as solid state alternative [45].

3   Fabrication Process of Nanostructures Used in Quantum Informatics

3.1   Fabrication of Quantum Dots

Nanotechnology and nanostructures are frequently used in the field of quantum informatics such as graphene, which have a structure of honeycomb network and is made of sp-hybridized carbon atoms [46, 47]. Graphene 2-D network has gained a lot of attention due to its astonishing properties like high electrical conductivity, high surface area, and mechanical stability [47]. It has also been depicted experimentally and theoretically that the morphology of graphene sheets characterizes its properties like size, shape including thickness [48–50].

Graphene quantum dots [49] having a size less than 100 nm and having properties like effective quantum confinement and edge properties can be used in optical and electronic devices. For the fabrication of graphene quantum dots [49]. Novoselov used electron beam lithography, but this fabrication technique is limited due to its low yield. Graphene quantum dots have also been prepared using graphene oxide sheets via different methods such as re-oxidization [51] and hydrothermal route [52, 53] and electrochemical avenue [51] with 10 nm diameter. All these methods involve top-down cutting process which gives no control over the size distribution of products.

Another way to achieve uniform graphene quantum dots is cyclo-dehydrogenation of poly-phenylene, but this method gives graphene quantum dots with size of 4 nm size that is below the processable scale. Uniformly shaped graphene quantum dots can also be obtained via cage opening of fullerene given by Loh [54]. Graphene and large poly cyclic aromatic hydrocarbons such as hexa-peri hexa-benzocoronene are nanoscale materials that have properties like high stacking tendency and high thermal stability [55]. By the pyrolysis of uniformed aromatic hydrocarbons, unprecedented carbon molecules were produced. The properties of these molecules can be produced via pyrolysis conditions [56–58].

We are going to present the fabrication of graphene quantum dots with diameter of 60 nm and thickness of 2–3 nm with mono-dispersed disk like shape. These graphene quantum dots are fabricated via the process of oxidization, carbonization, surface functionalization, and reduction with HBC precursors. These graphene quantum dots have the largest size and have PL efficiency of 3.8% suitable for nano devices.

The process to fabricate quantum dots is shown in Fig. 1.

The powder obtained from above-mentioned fabrication process presented well-defined scattering peaks in its WAXS pattern known as wide angle X-ray scattering. The WAXS pattern is given in Fig. 2.

For the preparation of artificial graphite as powder, it is pyrolysed at different temperatures such as 600, 900, and 1200 K for 5 h. The obtained products are named as AG-600, AG-900, and AG-1200 as shown in Fig. 2.

According to the WAXS pattern as given in Fig. 2, the artificial graphite temperature in thermal treatment is very important for nano disk architecture. With increased temperature, AG-1200 architecture as round nano disk is acheived. This shows that we can use nanostructures and nanotechnologies techniques to obtain different quantum dots that can be used as q-bits in certain applications of quantum informatics [60].

3.2 Fabrication of Josephson Junctions

Quantum computation has been intriguing scientists for many years to store useful information proficiently. It stores information in the form of q-bits. Quantum

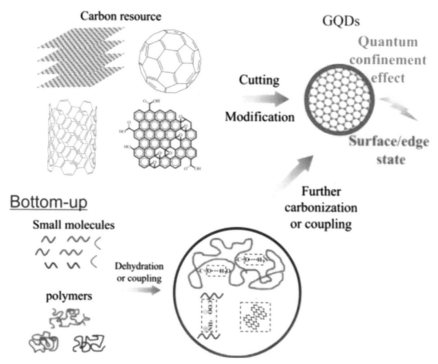

Fig. 1 Fabrication of photo-luminescent graphene quantum dots via bottom up technique with uniform morphology [59]

computers are faster than classical computers in functions such as factorization of large numbers and data mining. Systems such as NMR spectroscopy, ion trap, quantum electrodynamics have performed logic operations with fewer quantum bits. But to build a quantum computer using these systems is quite difficult [61]. Solid state systems that use Josephson junction-based semiconductor devices lose their coherence in a short time and strongly interact with their environment [62].

Maintaining coherence is one of the main problems of solid state quantum bit. In Josephson junction, the coherence is quality dependent [63]. The quality depends on the gap voltage, resistance, impedance, and properties of the material used. It is desirable to have low sub-gap leakage current [64, 65].

3.2.1 Electron Beam Exposure

UVN-30 with resolution of 0.18 μm is used as an electron beam resist. When silicon atoms are coated with this electron beam resist and are spun at the speed of 6000 rpm, it gives us a thickness of 400 nm. The resultant samples are heated for 1 min and

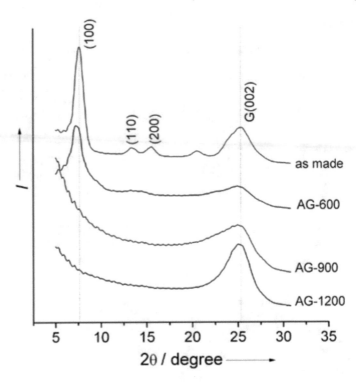

Fig. 2 WAXS pattern of as made powder for thermal treatment [60]

then exposed to electron beam of 30 kV potential. After the exposure of electron beam, the sample is heated again for another minute at 90 degree and is developed using CD-26 developer in DI water. Figure 3 shows different contrasts with different compositions of micro posit developer CD-26.

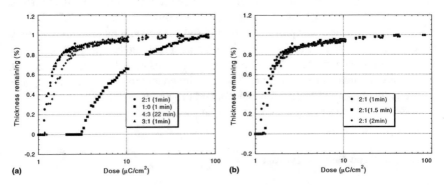

Fig. 3 Contrast curves of UVN-30 resist exposed with 30 kV electron beam and developed via micro posit CD-26 developer. **a** Contrast with 4:3, 4 parts of developer and 3 parts of DI, **b** contrast with 2:1 ratio [64]

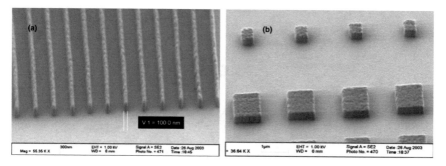

Fig. 4 **a** Single pass lines with size of 100 nm that are falling on UVN-30, **b** resist pattern of 1 and 0.5 μm [64]

With 1: 1, the development time is increased, but the contrast result is very poor. For 2:1 ratio of DI solution, the development time is longer, but there is no effect on contrast. The results obtained are given in Fig. 4.

3.2.2 Self-assigned Process

The devices are made on silicon wafers with diameter of 50 mm. The process of self-assigned process includes the steps as described below.

Firstly, a double layer of resist is bared through processes such as photolithography, electron beam lithography, and the resist that is used as a lift for electrode patterns. After that, a trilayer of NB/AlO$_x$/Nb is deposited in a chamber.

Figure 5 shows that firstly, a 150 nm thick layer of Nb is placed. After the deposition of Nb layer, an 8 nm thick layer of Al is deposited onto it. The Al layer oxidizes, and a layer of Nb is deposited again at base pressure. The first layer of Nb that was deposited is called the base electrode, and the second deposited layer of 150 nm thick Nb is called counter electrode. This trilayer is produced in a bath of acetone. In Josephson junction, the current density is controlled via oxygen in the process of oxidization. The counter electrode is used as beam and photon resist by using UVN-30 negative resist. In this counter electrode, 2:1 ratio of DI solution is used along with 5 μC/cm^2. Negative resist is used as an etch mask, and RIE is used to etch Nb in a SF^6 plasma.

Layer of quartz with thickness of about 150 nm is used as dielectric layer. This layer of dielectric is deposited via RF sputtering. The layer of deposited dielectric is defined via lift off performed by micro posit 1165. During the deposition of quartz, it is sometimes difficult to perform the lift off. To control this, it is desirable to control the sample cooling.

In the next step, contact holes are formed due to RIE of Nb and due to Al inter layer as an etch mask. Finally, thick Nb layer having a thickness of 200 nm defined via EBL is obtained (Fig. 6).

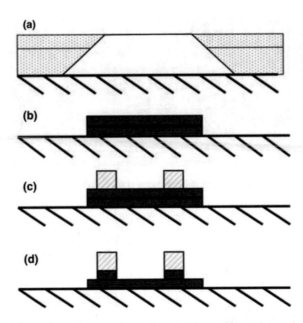

Fig. 5 **a** A double layer of resist is exposed through photolithography and electron beam lithography, and the resist is used as a lift for electrode patterns. **b** A trilayer of Nb/AlO$_x$/Nb is deposited in a chamber. **c** The counter electrode is used both as beam and photon resist by using UVN-30 negative resist. In this counter electrode, 2:1 of DI solution is used along with 5 μC/cm^2. **d** Negative resist is used as an etch mask, and RIE is used to etch Nb in a SF^6 plasma. Etch stop in this case is Al layer that naturally acts as etch mask [64]

Figure 7 shows etched Nb patterns. An under count occurs due to slightly over etching with a size of 60–70 nm. This helps in the lift off process. A SEM micrograph of a SQUID qubit that is used for quantum computer is shown in Fig. 7b, c.

4 The Role of Nanotechnology in Energy Solution

The key challenge in the present era is the advancement of sustainable power sources because of major concerns linked with the development and consumption of energy [66]. Nanotechnology has played a promising role in relieving the energy crisis problems in world and provided a wide range of their solutions in industries via new technologies. As it helped in production of new energy resources, energy conversion methods, energy distribution, storage and usage [67]. Sustainable power sources can be considered as a sort of vitality which can give light, power, and warmth without contaminating the earth. Energy from petroleum products has been characterized as major cause of environmental contamination. Preferred perspective of sustainable power sources is that no petrol is essential, which bestows with the discharge of carbon dioxide [68]. The utilization of oil utilization is 10^5 times quicker than the

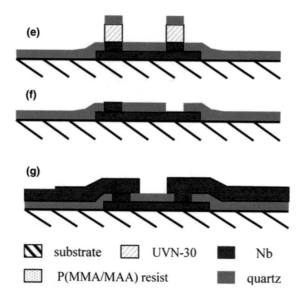

Fig. 6 **e** A 150 nm thick layer of quartz is used as dielectric layer. This layer of dielectric is deposited via RF sputtering. The layer of deposited dielectric is defined via a lift off performed by micro posit 1165. During the deposition of quartz, it is sometimes difficult to perform the lift off. To control this difficulty, it is desirable to control the sample cooling. **f** Contact holes are formed due to RIE of Nb and due to Al interlayer that is used an etch mask and uses PMMMA as resist, and these contact holes are called junctions initially. **g** In the final step of self-assigned process, a 200 nm thick Nb layer that is defined via EBL is obtained [64]

natural consumption, and as a consequence, it is predicted that world's petroleum storage will be endangered by 2050 [69].

4.1 Nanotechnology-Based Energy Resources

Nanotechnology provides a vital role in the production of both renewable and nonrenewable resources and has potential to solve all the problems related to Sun, water, biomass, wind and ocean waves, etc. Several nanomaterials are used to optimize the lifespan and efficiency of the system which produce energy through the deposition of oil and gas developments [70]. Some nanomaterials, which are in direct contact to the mechanical system of wind mills, are corrosion resistant for the system components that also control the fast wind blades and thus produce energy.

Numerous photovoltaic-materials are involved in regenerative process of energy resources like nano-optimized cells (some polymeric, multiple junctions, quantum dots and thin film dye, etc.) and coatings that are resistant to any kind of reflection. Wind energy comprises some nanocomposites as wind blades by increasing their

Fig. 7 a SEM micrograph of Nb and UVN-30 after RIE [64]. b A complete Josephson junction with SQUID qubit and required measurement magnetrons [64]. c A $0.2 \times 0.2 \ \mu cm^2$ Josephson junction [64]

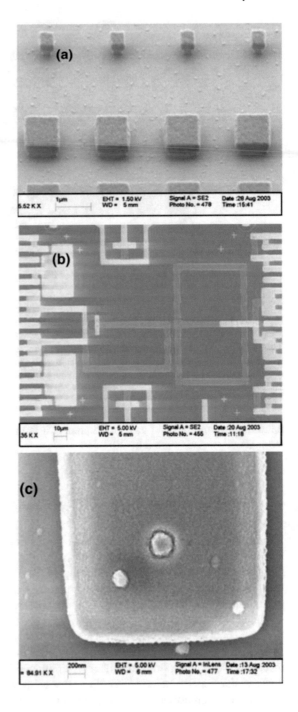

efficiency [71]. Geothermal energy involves some nanostructures and nanocomposites in digging and drilling processes [72]. Biomass energy involves a sufficient amount of energy by following some nano-based precisions and precautions related to forming and plantation. Some bio-nanosensors are also used in this process for optimization of energy. Fossil fuels in energy process involve some nanoparticles in drilling equipment. In nuclear field, nanosensors and nanocomposites are used for protection of fusion and radiation that are harmful for human health as well as for environment. Therefore, one can reduce these major problems in medical field by using some nanomaterials [67].

4.2 Energy Production Methods with Nanostructure

Nanotechnology predicts an extensive range of energy crisis solutions that how nanotechnology is used in different fields to produce energy [72].

4.2.1 Steam Generation

The surge of producing an increased level of energy to lessen the crisis and to meet the needs across the globe, some researchers depicted an increasing amount of energy that can be produced by falling a specific amount of sunlight on the nanomaterials which produced steam layer of high efficiency. Some researchers are trying to produce such nanoparticles that can produce steam for power plants also [73].

4.2.2 Highly Efficient Cost-Effective Bulbs

A polymer, made by nano engineering at nanoscale, is used to enhance the capacity of the bulbs by increasing the fluorescence capacity of the bulbs. These bulbs are more reliable than the other fluorescence bulbs as they are unbreakable and have double working capacity than others. Plasmonic cavities are used to increase the efficiency of the LEDs which are thin nano sixes array of blades. On the other end, incandescent bulbs can also be renewed by wrapping their filaments with some crystalline nanomaterials to increase the efficiency of these bulbs [74].

4.2.3 Fuel Cells

Nanotechnology is used in producing fuel cells in very efficient way by reducing the cost of catalysts (which are used in producing the hydrogen ions) used in these cells. Nanomaterials are used to create the membranes in cells which cause this catalyst to separate other ions like oxygen from hydrogen ions. Fuel in these cells can be produced from nanomaterials and uses the raw materials like diesel and gasoline (refined through several processes) [75]

Additional advantages of nanotechnology to cope with energy solutions are an increased amount of the energy generated through wind mills, production of electricity by extra heat, production of energy through storing of hydrogen for cars powered by fuel cells, some piezoelectric nano fiber cells that are good agents for generation of electricity, improvement in batteries by using nanomaterials, and reducing the cost of solar cells [76].

4.2.4 Geothermal Energy via Nanostructures

It is a type of energy which is extracted from the earth's crust. The temperature depends upon the depth. The higher the depth, the higher will be the temperature. In this matter, the nanotechnology plays a very vital role and helps to extract more energy. Nanofluids can be exploited as cooling reagents under particular temperature and friction conditions to cool down the channels and essential components like sensors and hardware in boring machines [77].

The geothermal gradient is a difference in temperature between earth's core and its surface. A constant thermal energy conduction drives from the center of the surface in the form of the heat. From the opposite side, nanofluids are used as working liquid to excerpt energy from the center of earth and controlled it into a power-plant framework to deliver a lot of energy [78]. Truth be told, geothermal-power plants can work 24 h for everyday, giving base-stack limit and the world-potential limit with respect to geothermal power which is assessed (85 GW) throughout the following 30-years [79]. Tremors and earthquakes are the major challenges which occur by utilizing the geothermal energy when deep boring is done. This perception unties a scope of future research forecasts of utilizing fewer profound borings to use geothermal vitality [80].

5 Role of Nanotechnology in Energy Solution

The energy is considered as vital for all human activities, but in present era, the energy consumption is greater than energy production. Nowadays, the major problem is the energy crises that can be minimized by the means of nanotechnology. Discussion about the potential effect of nanotechnology on the energy transmission, the solutions of the energy, and its practical applications is described in this section. In this article, it is discussed how nanotechnology reduces the energy crises which are useful in electricity field as batteries (most powerful source of energy transport). Nanomaterials can be utilized in various applications such as medicines, catalytic, solar cells, sensors, by exhibiting unique biological, physical, and chemical properties [81].

5.1 Power Losses

To lower the power loss in the wires, some carbon nanotubes are used in transmission wires that further lessen the resistance in the circuits. Richard Smalley intended in a conference about the benefits of nanotechnology. Using these concepts of Smalley, power losses can be reduced over long distances [82].

5.2 Impact on Energy Transmission

A carbon nanotube CNT is the type of nanostructure that consists of carbon molecules and formed when the atoms of the carbon are arranged in tubular shapes. CNT properties are like light and brilliant structure, best conductor than the copper and the steel and have high cost range. The carbon atoms are arranged in spherical array look like Bucky balls, used in many mechanical and semiconductor operations. Nano-dots and quantum dots are the nanostructures that have optical and the electrical properties, used in lighting and the solar collection. Another type of the CNTs the "Armchair CNTs" have more potential effect on the transmission of the energy than ordinary materials and best conductor like diamond, steel, and the copper, which conduct more electricity and reduced the energy losses. In this nanotube, the electrons move down like coherent wave in single mood optic fiber [73].

"Armchair CNTs" conduct and produce more current as ordinary CNT, when the current flow inward the tube due to the photons we called it "Armchair quantum wire" which can conduct the 20 A more current as ordinary CNTs providing the energy solution through this wire used in the tubes and the potential effect of the nanotechnology on the transmission of the energy can be reduced in this way which has many applications in the quantum nanotechnology [83].

6 Applications of Nanotechnology in Different Energy Development Fields

Nanotechnology serves in many applications and provides efficient way to minimize the potential effect and the transmission of the energy to make more perfect energy sources used in the electricity, distillation of the fuel, and the natural gas (purified through the nanotechnology). Nanotechnology produces more energy, makes the transmission of the energy easier in longer distance, maintains the energy, and minimizes the effect of the construction, repair, and the other activities along ROWs [70]. The energy-related fields in which the nanotechnology plays a vital role are described as given below.

- Lighting
- Heating

- Transport of the energy
- In renewable resources of the energy
- Storage of the energy
- Generation of the energy from the fuel cell
- Hydrogen energy generation and the storage of that energy
- Energy in power chips.

6.1 Lighting

In USA, almost 20% of electricity is consumed in providing light in terms of incandescent and fluorescent bulbs. The nanotechnology introduced LEDs, a semiconductor device that has the worthy characteristics like compactness, low heat generation, more duration, more electrical efficiency, and visible light spectrum (has more visibility due to shorter wavelength), used in many incandescent light bulbs, automobiles lights, and in traffic lights. The semiconductors which are the great invention of the nanotechnology, used in the formation of the LEDs for the lightening purpose and being built in nanoscale, show that the nanotechnology has potential to reduce the energy consumption and to minimize the potential effect of the surroundings on the transmission of the energy. LED is a thin layer of the quantum dots which produces the light of the different colors. The nanotechnology is used in the lighting at commercial and the industrial level [79].

6.2 Heating

CNT nano-fluid, formed by adding the water in CNT, is four times efficient than simple CNT in the process of transferring the heat. When this nanofluid is used in the commercial level as in water boiler at hoses level, then this nanofluid makes the boiler more efficient, and central heating device becomes 10% more efficient [84].

6.3 Transmission of the Energy

Nanotechnology offers much efficient way of transportation of the energy by using different techniques as diesel fuel additives. This method of the transportation of the energy reduces the fuel consumption and the concentration of harmful materials produced during the combustion of the fuel. At the nanoscale, the additives are used, as the particles of cerium oxide catalyze combustion reactions and the burning of the fuels in the presence of the air. The small particles provide the high surface area, the catalyst catalyzes the reaction for more combustion process to occur, and thus, energy production is more. Energy can be transported in efficient way by using the high strength materials owing to low weight that are designed on the basis of the nanotechnology (reduce the transportation of fuel that make the engines more efficient of the engines is 50% increased than the conventional engines). Nanotechnology works to improve the fuel economy for better transmission of the energy [85].

The capable batteries are created by using electrolytes composed of the nanomaterials. Lithium ion batteries are mostly used in nanotechnology which have the capability to increase the efficiency of the vehicles and engines in automobiles. The capacitors are made up of multi-walled central nanotube, which increase their energy storage capacity and make the capacitors more efficient. CNT tube dramatically affects the surface area of the capacitor by increasing it, as a consequence, more charge is produced and stored, and hence, this method reduces the energy transmission and is compact. The supercapacitors reduce the loss of energy in the vehicles and the automobiles engines, increase the efficiency of the hybrid electric vehicles, and reduced the consumption of fuel [86].

6.4 Energy Harvesting Using Piezoelectric Nanogenerators

Energy is produced, nowadays, by using different methods such as nuclear combustion, kinetic energy of water, and through geothermal sources. Nano electrochemical systems produce efficient and ambient amount of energy as it is needed in present age. Energy can be obtained through thermal gradients, solar energy, and piezoelectricity, a process which can convert the mechanical energy into electrical energy [87]. These nano devices work in such environment where certain variables need to be checked, so these are placed in office buildings and on the floor of oceans where humans cannot produce energy, and piezoelectric nanogenerators provide power to these nano devices [88].

Similarly, main sources of energy for human activities are minerals, kinetic energy from water, nuclear fission, hydroelectric power, and through thermal gradients, but these sources cause global warming, destruction of geosphere, ecological devastation, and depletion of ozone layer. Since all these factors produce environmental pollution that is harmful for humans as well as for animals [67]. About 80% of CO_2 is emitted in the air through industries. In contrast, clean and efficient amount of energy can be produced through renewable resources such as solar power, wind mills, and ocean tides. However, the energy production is limited due to its very high cost and limited resources [89]. For example, in USA, only 7% energy is produced by renewable resources, while 85% of energy is produced through fossil fuels and 8% through nuclear energy. Similarly, in Germany, major portion of energy is obtained from nuclear sources which can cause adverse effects on the health of people.

Nanotechnology, for the first time, developed such industries which are based on efficient and effective cost may help to sustain the energy resources in any country. It describes specific atoms and molecules to control the properties of materials. Since nanotechnology has reduced the size and cost of energy storing devices [90].

According to roadmap report, the concerning use of nanotechnology for production of energy is very challenging field especially in hydrogen conversion, thermal devices, photovoltaic devices, and solar energy.

6.5 Solar Economy

Solar economy deals with the processes that involve sunlight to produce energy in different ways. Solar energy is the cheapest way to obtain the energy, as it is easily available all around the world. Sun can produce almost 15,000 times more energy than other sources such as fossils and nuclear fission [91]. The energy sources in Sun are photovoltaic cells which convert light energy into electrical energy, artificial photosynthesis which produces carbohydrates or hydrogen through water splitting that's why called passive solar technology and solar thermal systems which are used in solar collectors. Even plants use solar radiations for completion of their chemical reactions which makes different carbohydrates to produce energy, steam and bio-fuels [92] (Fig. 8).

Biomass technologies are widely used for the production of energy. Biomass covers almost 9–13% of world's primarily energy resources. Biomass energy has low density as compared to solar energy and has less conversion efficiency but has capacity to store solar energy. Researchers are interested in the development of photoactive materials that directly convert sunlight to electricity and to produce solar thermal systems which stores radiations of solar energy [93].

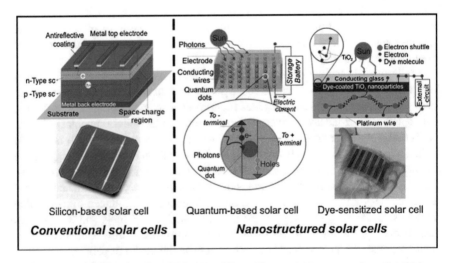

Fig. 8 From conventional solar cell based on silicon chip to nanostructures solar cells which are dye-sensitized cells based on quantum dots [67]

6.5.1 Photovoltaic Cells; for the Production of Electricity

Although, solar energy is free of cost and present in excess amount around the world but still, photovoltaic cell produces only 0.04% of total energy in the world. Recent advancements have reduced the prices of these methods to produce pollution free energy. Photovoltaic cells are like a device which works on photoelectric effect where photons convert the light energy into electric current to produce the electricity from radiations of Sun [94]. These photovoltaic cells are made up of silicon wafers having thickness of about 150–300 nm. This is the technology of first generation of photovoltaic cells. But the second generation of these cells is made of semiconductor thin films of thickness of about 1–2 nm and produces almost 86% of energy. Most of them are comprised of special epitaxial layers of semiconducting materials which cover almost 90% demand of market. In the field of nanotechnology, the solar cells are made of thin film conductive transparent oxide like indium tin oxide with the external coating of nanocrystals. The quantum dots are more efficient because they emit many electrons per solar photon with different emission and absorption properties in the spectrum based on the size of particle [95]. In 1991, the first solar cell named as dye-sensitized colloidal titanium oxide cell which is sandwiched between the layers of anode and cathode depends on the platinum anode which behaves as a carrier for the conduction and transportation of electrons. These dye-sensitized nanostructured devices are very commonly used in the market, and their efficiency of converting solar light to electricity can be increased by using quantum dots.

6.5.2 Artificial Photosynthesis; for Hydrogen Production

Photovoltaic cell has capability to split water into its constituents as hydrogen and oxygen which is known as water catalysis by photocatalytic process. This means that solar energy can directly be stored in the form of water via oxygen and hydrogen ions. Splitting of water involves different processes which converts water into hydrogen and oxygen. The hydrogen production can be made possible at high temperatures such as $1000°$ C due to thermochemical splitting of water with sulphur ionic oxide on carbon substrate. Later, the electrons are produced by providing the power to electron transport chain in process of photophosphorylation. Researchers suggest that hydrogen can be produced in a cheap way by splitting methane gas into hydrogen and carbon.

The splitting of water through photocatalytic process is known as artificial photo-synthesis. Nanotechnology produces cheap and efficient hydrogen from solar energy by using different nanostructured devices such as Cds, SiC, $CuInSe_2$, and TiO_2. Titanium oxide is more useful because it has bandgap of about 3.2 eV which means only UV light can be used for the water splitting and rapid electron hole pair recombination takes place, and we can increase the photocatalytic activity by two methods in which the first one is to add chemical substances to water and second is modification in photo catalyst. Nakato and its coworkers have synthesized a thin film electrode which is made of titanium oxide and doped with silicon, a semiconductor material which could absorb short as well as longer wavelengths that result in high conversion efficiency of solar energy to chemical energy to about 10% [96].

6.6 Hydrogen Economy

Hydrogen is a way of transporting and storing energy using natural methods such as thermal and wind powers. Hydrogen produces energy free of pollution because water is obtained through its combustion which has no harmful impacts. Therefore, combining production of hydrogen from renewable resources produces excess amount of environment-friendly energy.

6.6.1 Hydrogen Production and Storage

Nowadays, hydrogen is produced through steam reforming of natural gas, but the drawback of this process is that it produces 70–80% of CO_2 as its residual product which has harmful impacts [97]. Hydrogen can be produced by renewable resources such as biomass, bio-fuels, and electrolysis of water to produce electricity of about 70%. The available storage systems of hydrogen gas are quite insufficient because pressure vessels to store hydrogen are very heavy and to store hydrogen as a liquid requires very low temperature and density. Due to these shortcomings, most of the energy is wasted. Currently, physisorption and chemisorption are the most efficient ways to store this light gas [91].

6.7 Sustainable Energy Storage

Much of sustainable energy sources are required to store and transport energy. Capacitors and batteries are the examples of energy storage devices. There is a great contribution of nanotechnology to store the hydrogen gas [98].

6.7.1 Rechargeable Batteries

Rechargeable batteries based on lithium ions provide much more storage capacity than that of aqueous batteries. But there are some drawbacks such as low density and volume charge on the surface which can be accomplished through nanotechnology.

Sony has made tin rechargeable batteries made of electrodes named as Nexilion, the first alloy that replaces the graphite anodes. Toshiba made lithium ion rechargeable battery having a large storage capacity that can store 80% of energy in only 1 minute [99]. Electrolyte capacity can be increased by adding nanoparticles of alumina, zirconium, to liquid electrolyte of non-aqueous solution. These alloys increase the size of nanocomposites by reducing the volume change.

Another way to store energy is electrochemical capacitors also known as supercapacitors or ultra-capacitors which store electricity as rechargeable batteries but in a different way. It stores electricity by separating positive and negative charges. The conventional capacitors have many drawbacks such as high cost, low efficiency, and large requirements for their better life cycle. But the capacitors made of nanocomposites known as pseudo, redox supercapacitors, or electrochemically double layer capacitors have high efficiency. Transition of carbon electrodes to nanostructured carbon tubes using nano-templates (mesoporous, silica, and zeolites) provides high surface area and electrical conductivity [92].

7 Conclusion

This chapter accords with nanostructures and its role in quantum informatics, specifically quantum computation and energy solution. Nanostructures are of great deal of interest. Semiconductor nanostructures and their role in quantum computation are discussed. Quantum dots, as quantum bits have several advantages, have a disadvantage of small decoherence time. Quantum dots-based quantum computer is very attractive due to its entanglement properties. Quantum information processing is pondered with nanostructures such as solid state system. Scalability is an important factor in quantum information processing. It is suggested that quantum system with optically controlled interactions is more reliable because laser technology allows the quantum communication in the desired coherence time. Fabrication methods of nanostructures that are employed in quantum computation such as graphene quantum dots and Josephson junction are also reviewed with specific techniques.

Nanostructures can be employed both in renewable and nonrenewable energy sources. Some nano-coated materials are used to optimize the lifespan and efficiency of the system which produce energy through the deposition of oil and gas developments.

References

1. Nalwa, H.S.: Handbook of Nanostructured Materials and Nanotechnology, Five-Volume Set. Academic, London (1999)
2. Luryi, S., Xu, J., Zaslavsky, A.: Future Trends in Microelectronics: The Nano Millennium. Wiley-IEEE Press, New York (2002)
3. Nag, B.R.: Physics of Quantum Well Devices. Springer Science & Business Media, Berlin (2001)
4. Herman, M.A., Sitter, H.: Molecular Beam Epitaxy: Fundamentals and Current Status. Springer Science & Business Media, Berlin (2012)
5. Lieber, C.M., Wang, Z.L.: Functional nanowires. MRS Bull. **32**(2), 99–108 (2007)

6. Xia, Y., Yang, P., Sun, Y., Wu, Y., Mayers, B., Gates, B., Yin, Y., Kim, F., Yan, H.: One-dimensional nanostructures: synthesis, characterization, and applications. Adv. Mater. **15**(5), 353–389 (2003)

7. Logothetidis, S.: Nanostructured Materials and Their Applications. Springer Science & Business Media, Berlin (2012)

8. Nakamura, S., Senoh, M., Iwasa, N., Nagahama, S.I.: High-brightness InGaN blue, green and yellow light-emitting diodes with quantum well structures. Jap. J. Appl. Phys. **34**(7A), L797 (1995)

9. Andersson, J., Lundqvist, L.: Near-unity quantum efficiency of AlGaAs/GaAs quantum well infrared detectors using a waveguide with a doubly periodic grating coupler. Appl. Phys. Lett. **59**(7), 857–859 (1991)

10. Petroff, P.M., Lorke, A., Imamoglu, A.: Epitaxially self-assembled quantum dots. Phys. Today Am. Inst. Phys. **54**(5), 46–52 (2001)

11. Kirstaedter, N., Ledentsov, N., Grundmann, M., Bimberg, D., Ustinov, V., Ruvimov, S., Maximov, M., Kop'ev, P.S., Alferov, Z.I., Richter, U.: Low threshold, large T/sub o/injection laser emission from (InGa)As quantum dots. Electron. Lett. **30**(17), 1416–1417 (1994)

12. Loss, D., Divincenzo, D.P.: Quantum computation with quantum dots. Phys. Rev. A **57**(1), 120

13. Divincenzo, D.P.: Topics in Quantum Computers. Springer, Dordrecht (1996)

14. Hanson, R., Witkamp, B., Vandersypen, L., Van Beveren, L. W., Elzerman, J., Kouwenhoven, L.P.: Zeeman energy and spin relaxation in a one-electron quantum dot. Phys. Rev. Lett. **91**(19), 196802 (2003)

15. Bednarek, S., Szafran, B., Adamowski, J.: Theoretical description of electronic properties of vertical gated quantum dots. Phys. Rev. Lett. **64**(19), 195303 (2001)

16. Nielsen, M.A., Chuang, I.: Quantum computation and quantum information. Am. J. Phys. (2002)

17. Ekert, A., Jozsa, R.: Quantum computation and Shor's factoring algorithm. Rev. Mod. Phys. **68**(3), 733 (1996)

18. Shor, P.W.: Polynomial-time algorithms for prime factorization and discrete logarithms on a quantum computer. SIAM Rev. **41**(2), 303–332

19. Shor, P.W.: Scheme for reducing decoherence in quantum computer memory. Phys. Rev. A **52**(4), R2493 (1995)

20. Kane, B.E.: A silicon-based nuclear spin quantum computer. Nature **393**(6681), 133 (1998)

21. Imamog, A., Awschalom, D.D., Burkard, G., Divincenzo, D.P., Loss, D., Sherwin, M., Small, A.: Quantum information processing using quantum dot spins and cavity QED. Phys. Rev. Lett. **83**(20), 4204 (1999)

22. Žutić, I., Fabian, J., Sarma, S.D.: Spintronics: fundamentals and applications. Rev. Mod. Phys. **76**(2), 323 (2004)

23. Sarma, S. D., Fabian, J., Hu, X., ŽUtić, I.: Spin electronics and spin computation. Solid State Commun. **119**(4–5), 207–215 (2001)

24. Voyles, P., Muller, D., Grazul, J., Citrin, P., Gossmann, H.J.L.: Atomic-scale imaging of individual dopant atoms and clusters in highly n-type bulk Si. Nature **416**(6883), 826 (2002)

25. Grabert, H., Devoret, M.H.: Single Charge Tunneling: Coulomb Blockade Phenomena in Nanostructures, vol. 294. Springer Science & Business Media, Berlin (2013)

26. Dykman, M., Santos, L., Shapiro, M., Izrailev, F.: Many-particle localization by constructed disorder and quantum computing. In: AIP Conference Proceedings, pp.148–159. AIP (2005)

27. De Sousa, R., Delgado, J., Sarma, S.D.: Silicon quantum computation based on magnetic dipolar coupling. Phys. Rev. A **70**(5), 052304 (2004)

28. Hu, X., Sarma, S.D.: Gate errors in solid-state quantum-computer architectures. Phys. Rev. A **66**(1), 012312 (2002)

29. Sarma, S.D., De Sousa, R., Hu, X., Koiller, B.: Spin quantum computation in silicon nanostructures. Solid State Commun. **133**(11), 737–746

30. Bennett, C.H., Shor, P.W.: Quantum information theory. IEEE Trans. Inf. Theory **44**(6), 2724–2742 (1998)

31. Divincenzo, D.P.: The physical implementation of quantum computation. Fortschritte Der Physik Progr. Phys. **48**(9–11), 771–778 (2000)
32. Monroe, C.: Quantum information processing with atoms and photons. Nature **416**(6877), 238 (2002)
33. Eschner, J., Blatt, R.: Realization of the Cirac–Zoller controlled-NOT quantum gate. Nature **422**(6930), 408 (2003)
34. Gulde, S., Riebe, M., Lancaster, G. P., Becher, C., Eschner, J., Häffner, H., Schmidt-Kaler, F., Chuang, I. L., Blatt, R.: Implementation of the Deutsch–Jozsa algorithm on an ion-trap quantum computer. Nature **421**(6918), 48 (2003)
35. Loss, D., Divincenzo, D.P.: Quantum computation with quantum dots. Phys. Rev. A **57**(1), 120–126 (1998)
36. Kane, B.E.: Silicon-based quantum computation. Fortschritte der Physik Progr. Phys. **48**(9–11), 1023–1041 (2000)
37. Quiroga, L., Johnson, N.F.: Entangled bell and Greenberger-Horne-Zeilinger states of excitons in coupled quantum dots. Phys. Rev. Lett. **83**(11), 2270 (1999)
38. Biolatti, E., Iotti, R.C.,, Zanardi, P., Rossi, F.: Quantum information processing with semiconductor macroatoms. Phys. Rev. Lett. **85**, 5647 (2000)
39. Platzman, P., Dykman, M.I.: Quantum computing with electrons floating on liquid helium. Scinece **284**(5422), 1967–1969 (1999)
40. Bimberg, D., Grundmann, M., Ledentsov, N.: Quantum Dot Heterostructures. Wiley, New York (1999)
41. Cooper, J., Valavanis, A., Ikonić, Z., Harrison, P., Cunningham, J.: Finite difference method for solving the Schrödinger equation with band nonparabolicity in mid-infrared quantum cascade lasers. J. Appl. Phys. **108**(11), 113109 (2010)
42. Burkard, G., Loss, D., Divincenzo, D.P.: Coupled quantum dots as quantum gates. Phys. Rev. B **59**(3), 2070 (1999)
43. Shah, J.: Ultrafast Spectroscopy of Semiconductors and Semiconductor Nanostructures, vol. 115. Springer Science & Business Media, Berlin (2013)
44. Kiraz, A. Reese, C., Gayral, B., Zhang, L.,Schoenfeld, W.V., Gerardot,B.D., Petroff,P.M., Hu, EL., Imamoglu, A.: Cavity-quantum electrodynamicss with quantum dots. J. Opt. B Quantum Semiclassical Optics **5**, 129 (2003)
45. Bonadeo, N.H., Erland, J., Gammon, D., Park, D., Katzer, D., Steel, D.: Coherent optical control of the quantum state of a single quantum dot. Science **282**(5393), 1473–1476 (1998)
46. Yavari, F.: Graphene nano-devices and nano-composites for structural, thermal and sensing applications. Rensselaer Polytechnic Institute (2012)
47. Schedin, F., Geim, A., Morozov, S., Hill, E., Blake, P., Katsnelson, M., Novoselov, K.S.: Detection of individual gas molecules adsorbed on graphene. Naturematerials **319**(5867), 1229–1232
48. Li, X., Wang, X., Zhang, L., Lee, S., Dai, H.: Chemically derived, ultrasmooth graphene nanoribbon semiconductors. Science **319**(5867), 1229–1232 (2008)
49. Ponomarenko, L., Schedin, F., Katsnelson, M., Yang, R., Hill, E., Novoselov, K., Geim, A.: Chaotic dirac billiard in graphene quantum dots. Science **320**(5874), 356–358
50. Kosynkin, D.V., Higginbotham, A.L., Sinitskii, A., Lomeda, J.R., Dimiev, A., Price, B.K., Tour, J.M.: Longitudinal unzipping of carbon nanotubes to form graphene nanoribbons. Nature **458**(7240), 872 (2009)
51. Shen, J., Zhu, Y., Chen, C., Yang, X., Li, C.: Facile preparation and upconversion luminescence of graphene quantum dots. Chem. Comm. **47**(9): 2580–2582 (2011)
52. Pan, D., Zhang, J., Li, Z., Wu, M.: Hydrothermal route for cutting graphene sheets into blue-luminescent graphene quantum dots. Adv. Matter. **22**(6), 734–738 (2010)
53. Gupta, V., Chaudhary, N., Srivastava, R., Sharma, G.D., Bhardwaj, R., Chand, S.: Luminscent graphene quantum dots for organic photovoltaic devices. J. Am. Chem. Soc. **133**(26), 9960–9963 (2011)
54. Lu, J., Yeo, P.S.E., Gan, C.K., Wu, P., Loh, K.P.: Transforming C 60 molecules into graphene quantum dots. Nat. Nanotechnol. **6**(4), 247–252 (2011)

55. Wu, J., Pisula, W., Müllen, K.: Graphenes as potential material for electronics. Chem. Rev. **107**(3), 718–747 (2007)
56. Zhi, L., Müllen, K.: A bottom-up approach from molecular nanographenes to unconventional carbon materials. J. Mater. Chem. **18**(13), 1472–1484 (2008)
57. Wang, X., Zhi, L., Tsao, N., Tomović, Ž., Li, J., Müllen, K.: Transparent carbon films as electrodes in organic solar cells. Angew. Chem. Int. Ed. **47**(16), 2990–2992 (2008)
58. Liu, R., Wu, D., Feng, X., Müllen, K.: Nitrogen-doped ordered mesoporous graphitic arrays with high electrocatalytic activity for oxygen reduction. Angew. Chem. **122**(14), 2619–2623 (2010)
59. Zhu, S., Song, Y., Wang, J., Wan, H., Zhang, Y., Ning, Y., Yang, B.: Photoluminescence mechanism in graphene quantum dots: quantum confinement effect and surface/edge state. Nano Today **13**, 10–14 (2017)
60. Liu, R., Wu, D., Feng, X., Müllen, K.: Bottom-up fabrication of photoluminescent graphene quantum dots with uniform morphology. J. Am. Chem. Soc. **133**(39), 15221–15223 (2011)
61. Chiorescu, I., Nakamura, Y., Harmans, C. M., Mooij, J.E.: Coherent quantum dynamics of a superconducting flux qubit. Science **299**(5614), 1869–1871 (2003)
62. Pashkin, Y.A., Yamamoto, T., Astafiev, O., Nakamura, Y., Averin, D.V., Tsai, J.S.: Quantum oscillations in two coupled charge qubits. Nature **421**(6925), 23–826 (2003)
63. Friedman, J.R., Patel, V., Chen, W., Tolpygo, S. K., Lukens, J.E.: Quantum superposition of distinct macroscopic states. Nature **406**(6791), 43–46 (2000)
64. Chen, W., Patel, V., Lukens, J.E.: Fabrication of high-quality Josephson junctions for quantum computation using a self-aligned process. Microelectron. Eng. **73**, 767–772 (2004)
65. Chen, W., Rylyakov, A.V., Patel, V., Lukens, J.E., Likharev, K.K.: Rapid single flux quantum T-flip flop operating up to 770 GHz. IEEE Trans. Appl. Supercond. **9**(2), 3212–3215 (1999)
66. Hussein, A.: Reviews, applications of nanotechnology in renewable energies—a comprehensive overview and understanding. Renew. Sust. Energ. Rev. **42**, 460–476 (2015)
67. Serrano, E., Rus, G., Garcia-Martinez, J.: Nanotechnology for sustainable energy. Renew. Sustain. Energy Rev. **13**(9), 2373–2384 (2009)
68. Satyanarayana, K.G., Mariano, A.B., Vargas, J.V.C.: A review on microalgae, a versatile source for sustainable energy and materials. Int. J. Energy Res. **35**(4), 291–311 (2011)
69. Demirbas, A.: Global renewable energy projections. Energy Sour. Part B **4**(2), 212–224 (2009)
70. Aithal, P., Aithal, S.: Ideal technology concept & its realization opportunity using nanotechnology. IJAIEM **4**(2), 153–164
71. Moniz, E.J.: Nanotechnology for the Energy Challenge. Wiley, New York (2010)
72. Wang, Z.L., Wu, W.: Nanotechnology-enabled energy harvesting for self-powered micro-/nanosystems. Angew. Chem. Int. Ed. **51**(47), 11700–11721 (2012)
73. Zäch, M., Hägglund, C., Chakarov, D., Kasemo, B.: Nanoscience and nanotechnology for advanced energy systems. Curr. Opin. Solid State Mater. Sci. **10**(3–4), 132–143 (2006)
74. Diallo, M.S., Fromer, N.A., Jhon, M.S.: Nanotechnology for sustainable development: retrospective and outlook. In: Nanotechnology for Sustainable Development, pp. 1–16. Springer, Berlin (2013)
75. Chen, Y.X., Lavacchi, A., Miller, H.A., Bevilacqua, M., Filippi, J., Innocenti, M., Marchionni, A., Oberhauser, W., Wang, L., Vizza, F.: Nanotechnology makes biomass electrolysis more energy efficient than water electrolysis. Nature Commun. **5**, 4036 (2014)
76. Ellingsen, L.A.-W., Hung, C.R., Majeau-Bettez, G., Singh, B., Chen, Z., Whittingham, M.S., Strømman, A.H.: Nanotechnology for environmentally sustainable electromobility. Nature Nanotechnol. **11**(12), 1039 (2016)
77. Kalogirou, S.A.: Seawater desalination using renewable energy sources. Progress Energy Combustion Sci. **31**(3), 242–281 (2005)
78. Pahl, G.: Power from the People: How to Organize, Finance, and Launch Local Energy Projects. Chelsea Green Publishing (2012)
79. Diallo, M., Brinker C.J.: Nanotechnology for sustainability: environment, water, food, minerals, and climate. In: Nanotechnology Research Directions for Societal Needs in 2020, pp. 221–259. Springer, Berlin (2011)

80. Ganguly, S., Banerjee, D., Kargupta, K.: Nanotechnology and nanomaterials for new and sustainable energy engineering. In: Proceedings of the International Conference Nanomaterials: Applications and Properties. Sumy State University Publishing (2012)

81. Guo, K.W.: Green nanotechnology of trends in future energy: a review. Int. J. Energy Res. **36**(1), 1–17 (2012)

82. Abdin, Z., Alim, M.A., Saidur, R., Islam, M.R., Rashmi, W., Mekhilef, S., Wadi, A.: Solar energy harvesting with the application of nanotechnology. Renew. Sustain. Energy Rev. **26**, 837–852 (2013)

83. Roco, M.C.: The long view of nanotechnology development: the National Nanotechnology Initiative at 10 years. Springer (2011)

84. Jones, R.: Nanotechnology, energy and markets. Nat. Nanotechnol. **4**(2), 75 (2009)

85. Han, J., Yang, Y.Z., Wang, S.Y., Luo, C.H., Xia, C.B.: Novel synthesis of semiconductor nanocrystalline ZnS. Mater. Sci. Technol. **23**(2), 250–251 (2007)

86. Rabiee, M., Mirhabibi, A.R., Zadeh, F.M., Aghababazadeh, R., Pour, E.M., Lin, L.: A new method of biomolecular recognition of avidin by light scattering of ZnS:Mn nano-particles. Pigment Resin Technol. (2008)

87. Kumar, B., Kim, S.W.: Energy harvesting based on semiconducting piezoelectric ZnO nanostructures. Nano Energy **1**(3), 342–355 (2012)

88. Hu, X., Li, G., Yu, J.C.: Design, fabrication, and modification of nanostructured semiconductor materials for environmental and energy applications. Langmuir **26**(5), 3031–3039 (2009)

89. Lai, X., Halpert, J.E., Wang, D.: Recent advances in micro-/nano-structured hollow spheres for energy applications: from simple to complex systems. Energy Environ. Sci. **5**(2), 5604–5618 (2012)

90. Kramer, I.J., Sargent E.H.: Colloidal quantum dot photovoltaics: a path forward. ACS Nano **5**(11), 8506–8514 (2011)

91. Stirling, D.A.: Nanotechnology applications. In: The Nanotechnology Revolution, pp. 281–434. Pan Stanford (2018)

92. Gampala, P., Nagalakshmi, M.B., Mahalakshmi, V.J.N.: Nano Engineering of Materials and Surfaces. **4**(8) (2018)

93. Filippin, A.N., Sanchez-Valencia, J.R., Garcia-Casas, X., Lopez-Flores, V., Macias-Montero, M., Frutos, F., Borras, A.: 3D core-multishell piezoelectric nanogenerators. Nano Energy **58**, 476–483 (2019)

94. Tong, C.: Materials-based solutions to advanced energy systems. In: Introduction to Materials for Advanced Energy Systems, pp. 1–86. Springer, Berlin (2019)

95. Rajbongshi, B.M., Verma, A.: Emerging Nanotechnology for third generation photovoltaic cells. In: Nanotechnology: Applications in Energy, Drug and Food, pp. 99–133. Springer, Berlin

96. Tong, C.: Perspectives and future trends. In: Introduction to Materials for Advanced Energy Systems, pp. 819–891. Springer, Berlin (2019)

97. Garbayo, I., Baiutti, F., Morata, A., Tarancon, A.: Engineering mass transport properties in oxide ionic and mixed ionic-electronic thin film ceramic conductors for energy applications. J. Eur. Ceram. Soc. **39**(2–3), 101–114 (2019)

98. Palit, S., Hussain, C.M.: Recent advances in green nanotechnology and the vision for the future. In: Green Metal Nanoparticles: Synthesis, Characterization and their Applications, pp. 1–21 (2018)

99. Centi, G., Čejka, J.: Needs and Gaps for Catalysis in Addressing Transitions in Chemistry and Energy from a Sustainability Perspective (2019)

Production of Bioplastics by Different Methods—A Step Toward Green Economy: A Review

Mujahid Farid, Kashaf Ul Khair, Sana Bakht, Warda Azhar, Muhammad Bilal Shakoor, Muhammad Zubair, Muhammad Rizwan, Sheharyaar Farid, Hafiz Khuzama Ishaq, and Shafaqat Ali

Abstract The human dependence on plastic has been recognized for decades and is considered as handy and useful material for many aspects of the essential requirements. A wide range of benefits of the plastic determines its lasting stability, resistant to reactions, anticorrosive and waterproofing applications. Among the worst, the plastic bags have most detrimental effects on marine creatures and recent studies found a huge concentration of microplastic in ocean depths. On the basis of emerging environmental concerns, it is the need of the hour to produce such sort of plastic material which can easily be broken down under natural environmental condition. This review paper focuses mainly on the methods being used for the production of biodegradable plastics along with their degradation processes for researchers and stockholders for future orientations. After degradation, the impact of bioplastics on the environment, fate of bioplastics in marine and land ecosystem has also been discussed. The last section of this study comprises the discussion that how biodegradable plastics are beneficial for our environment, economy and society. Besides other concerns of bioplastics, different obstacles prevail in the use of bioplastics are also

Authors' Contribution: All authors equally contributed in the present chapter.

M. Farid (✉) · K. U. Khair · S. Bakht · W. Azhar · H. K. Ishaq
Department of Environmental Sciences, University of Gujrat, Hafiz Hayat Campus, Gujrat 50700, Pakistan

M. B. Shakoor
College of Earth and Environmental Sciences, University of the Punjab, Lahore 54000, Pakistan

M. Zubair
Department of Chemistry, University of Gujrat, Hafiz Hayat Campus, Gujrat 50700, Pakistan

M. Rizwan · S. Ali
Department of Environmental Sciences and Engineering, Government College University, Faisalabad 38000, Pakistan

S. Farid
Department of Biology, Ecology and Evolution, Faculty of Sciences, University of Liege, Liege, Belgium

S. Ali
Department of Biological Sciences and Technology, China Medical University, Taichung 40402, Taiwan

109

studied along with advantages of biodegradable plastics as natural substances which do not require separate collection, sorting, recycling or any other waste solution (disposal at landfills or burning) as is the case with non-biodegradable plastics. So, it is well evident that production and use of bioplastics will prove to be a great step for developing green economy.

Keywords Biodegradability · Bioplastics · Green economy · Microbial synthesis · Waste materials

Abbreviations

CDA	Cellulose diacetate
GHG	Greenhouse gases
MCC	Microcrystalline cellulose
PBAT	Polyesteramides, aliphatic
PBSA	Aromatic co-polyesters
PE	Polyethylene
PHAs	Polyhydroxyalkanoates
PHB	Polyhydroxybutyrate
PHBCAB	Polyhydroxybutyrate synthesis genes
PLA	Polylactic acid fiber
SBR	Sequencing batch reactor
TFA	Trifluoroacetic acid
TGA/DSC	A Mettler Toledo thermogravimetric/differential scanning calorimetric
UV	Ultraviolet

1 Introduction

Plastic is one of the useful and greatest inventions ever introduced to mankind which brought revolution in every field of life. It was invented in the mid of the 1800s. It was the era during which all the scientists of the world were performing research and experimentation on a substance named as rubber. An English metallurgist named as Alexander Parkes was the first person who prepared plastic in the laboratory for the first time. Plastics are artificially synthesized polymers which are not biodegradable and not naturally broken down by microorganisms. On the basis of performance and properties, plastics have attained much importance than materials which are made up of wood and metal [1]. According to an estimated report, the annual production of plastics in the world is more than 300 million tons in 2015 [2]. The reason behind this huge production is its wide use and provides many versatile applications in many fields [1]. If we discuss the problems which are associated with the excessive use of plastic than its landfill problems which causes environmental pollution. Excessive usage of plastic products leads to the excessive generation of solid waste, so management of this huge amount of waste requires extensive costs and strong legislation. Moreover, landfilling capacity is shrinking due to extensive landfilling around the globe [3]. Secondly, accumulation of plastic waste in oceans which not only causes the pollution of marine ecosystems but is also harmful for marine species like fishes because of their ingestion of plastic fragments which clogs the fish gills and blocks the digestive tract ultimately leads to death. A study conducted on North Atlantic Ocean revealed that 580,000 pieces of plastic are present per square kilometers in one seawater sample [4]. The third problem is the plastic incineration, for example, burning of plastic at a high-temperature condition which leads to the emission of toxic greenhouse gases (GHGs) in the atmosphere such as methane and carbon dioxide. These gasses have the ability to trap heat and cause global warming leading toward climate change. The fourth major problem is the non-biodegradability of plastic and hence persists in the environment for hundreds of years [5]. In the past, it was thought that two-thirds of the earth surface is covered with oceans and this much quantity of water is enough to dilute all kind of pollution thrown in it. But now this approach of thinking is proving to be untrue because at the current situation the oceans have become polluted to a dangerous extent. Due to this, many sea creatures including a variety of fishes have gone extinct. It is also observed after examination that many fishes which were caught from the Atlantic Ocean are not suitable for eating. Beautiful harbors and bays where fishes and other aquatic life use to thrive have now turned into dirty and uninhabitable places. The main culprit of all these problems is the plastic.

At present, the emerging environmental, social and economic problems have triggered mankind to develop environmental-friendly bioplastic [6]. These are the novel products of the twenty-first century and of great importance, especially because of their biodegradable nature. Bioplastics are defined as biopolymers produced from natural renewable resources, and they are non-toxic and biodegradable. They are produced by biological systems which include plants, animals and microorganisms or can be produced chemically by biological materials like starch, sugars, natural fats, etc. Two strategies are involved in the conversion of raw materials into bioplastics. The first step is the extraction of required polymer from animal tissue or plant, and second is the polymerization of monomer through biotechnological or chemical route [7].

There is a variety of different sources and materials for the production of bioplastics which occur naturally. Agricultural feedstock, animals, microbial community and marine industrial waste are the four major sources which provide natural raw materials for the formation of bioplastics. These include cellulose, sugarcane, plant oil, hemp, weeds, corn starch and potato starch [8] (Fig. 1).

Biodegradable plastics prove to be the best alternative of conventional plastics (petroleum-based polymers) because they can degrade easily in natural environmental conditions. Degradation time of bioplastics is about a few weeks, while others may require several years to degrade. When bioplastics are disposed off in environmental systems, they are broken down into simpler forms by the enzymatic activity of

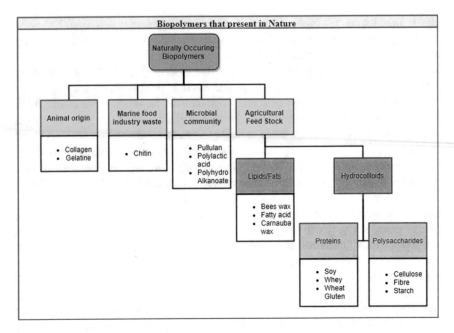

Fig. 1 Naturally occurring biopolymers and their sources [9]

bacteria, fungi and/or algae [10]. However, these polymers are also degraded by non-enzymatic processes like chemical hydrolysis. Keeping in view all above-mentioned environmental concerns, the process of replacing the plastics with the bioplastics has become a necessity of time.

Focusing on current environmental concerns of plastic pollution and its harmful impacts, it is the need of the hour to introduce such type of bioplastic which can degrade easily in the natural environment [11]. So, this review paper focuses mainly on the methods which can be used for the formation of biodegradable plastics. The most important objectives of this review paper include the following:

i.   To provide different production methods of plastics on one platform
ii.   To create awareness about the benefits of bioplastics in the growth of the green economy
iii. To provide a comparison between conventional plastics and biodegradable plastics
iv.   To provide value addition of waste recycling and its credibility and importance
v.   To fulfill sustainable development goals no. 7 and 11.

By fulfilling the above-mentioned objectives, this review paper proves to be very helpful, as it discusses the alternative of conventional plastics which create a lot of pollution and also persist in our environment for hundreds of years. It discusses different methods of production of environmentally friendly bioplastics. In this way, it could play a vital role in creating awareness about the detrimental effects of plastic on the environment as well as to combat the issue of plastic pollution by the use of bioplastics.

2 Classification of Biodegradable Plastics

There are a variety of sources and materials for the production of bioplastics including cellulose, sugarcane, vegetable oils, hemp, and corn and potato starch. But the classification of biodegradable plastic is a bit different. There are two main groups and four major categories. The first group is falling into argo-polymers which includes proteins and polysaccharides, and the second group bears biopolymers of polyhydroxyalkanoate, polylactic acid, and aliphatic and aromatic co-polyesters [8].

The biodegradable plastics can be categorized into four major categories on the base of material of production [12] (Fig. 2).

(a) Polymers based on biomass, for example, the agro-polymers from agro-materials. These are divided into two classes:

    i. Polysaccharides include lingo-cellulosic materials (straw and wood) [13], starches (maize, potatoes and wheat) and others like gums, chitin and pectin [14].

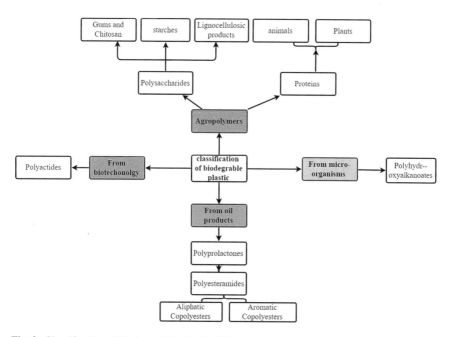

Fig. 2 Classification of biodegradable plastics [8]

ii. Lipids and proteins consist of plants such as gluten, zein and soya [15] and extract raw material from animals such as whey, casein and gelatin [16].

(b) Polymers which are obtained from agro-materials and chemically synthesized using monomers, for example, polylactic acid.

(c) Polymers gained by microbial synthesis and production. Its examples are polyhydroxyalkanoate and polyhydroxybutyrate.

(d) Polymers which are obtained from fossil assets by using chemical processes. Polyesteramides, aliphatic (PBAT) and aromatic co-polyesters (PBSA) are its examples [17].

The last category is the only one which is obtained from non-renewable materials. The first category is known as agro-polymers, and other are considered as bio-polyesters [15]. These biodegradable plastics are playing a very beneficial role in the improvement of surgery and food packaging industries and provide help to reduce carbon dioxide emissions and carbon footprint [18].

3 Biodegradation Stages of Bioplastics

Specific plastic materials have the property of biodegradation due to the polymeric nature [19].

3.1 Biodegradation

Biodegradation is basically the breakdown of complex compounds by microorganisms [20]. There are two types of biodegradation which include aerobic and anaerobic biodegradation. Aerobic biodegradation means the breakdown of polymers in the presence of oxygen by microorganisms carried out by aerobic bacteria because their metabolism is oxygen dependent. It results in complete degradation without producing pungent smell [21]. Anaerobic biodegradation is a process in which the breakdown of the polymer occurs in the absence of oxygen by anaerobic bacteria. First, the polymers broke down to small molecules so that making them available for other bacteria [10]. There are a number of bacteria involved in the process of anaerobic biodegradation like acetogenic bacteria convert these polymers into sugars, amino acids, hydrogen and organic acid ammonia and then methanogen bacteria convert them finally into methane and carbon dioxide [22].

Many methods of degradation of bioplastics are reported in the literature. One of the most important ways is by microorganisms like bacteria, fungi and yeasts which release different enzymes. This process is carried out in three steps.

- **Fragmentation** The first step is of fragmentation which mostly occurs outside of the organism and depends on the size of the polymer [23].
- **Biodegradation or Mineralization** The second step is biodegradation or mineralization carried out by microbes [24].
- **Assimilation** The third step is of assimilation of biodegraded polymers.

3.2 Bio-fragmentation

Generally, decomposition starts with the fragmentation. In this, chemical decomposition of polymer occurs on exposing with the living or non-living factors so that mechanical decomposition of the polymer occurs. The process of fragmentation proceeds due to a combination of thermal oxidation, ultraviolet (UV) radiations and also by microbial activity. UV radiations cause cracking and making fragments which leads to microplastics [25]. So, this means higher fragmentation occurs on exposure to UV radiations along with high oxygen level and a high temperature will boost up the fragmentation. During the process of fragmentation, it is covered in sediments and other organic or inorganic layers so the decrease in fragmentation process occurs rapidly. Addition of thermal and UV stabilizers can stop the process of fragmentation [26].

3.3 Bio-mineralization

A decomposition process in which complex organic substances are broken down (e.g., polymers) into smaller inorganic substances such as carbon dioxide takes place. In this way, plastics are converted into natural substances [27].

3.4 Bio-assimilation

Bio-assimilation refers to the process of providing nutrients to animal cells in a combination of two steps. In the first step, vitamins, chemical substances and minerals are absorbed from food which was converted into a simple form like carbon dioxide and sugar, while in the second step substances present in the bloodstream are chemically altered by the liver.

In the environment, everything has a good and bad impact, they fall in the white or black area based on their characteristics but bioplastics are included in a gray area because of many issues based on their benefits and problems. The biodegradation of bioplastics depends upon their chemical and physical structure therefore in different environmental mediums they possess a unique role in biodegradation. Different bacteria and fungus species play their part in biodegradation of bioplastics, and it depends on the type of bioplastics and the environment in which they are present. Biodegradation of bioplastics depends on different factors such as accumulation in the different environmental conditions and how much they are degraded by different microorganisms [28].

The fate of bioplastics in soil and marine environment is given as follows:

3.4.1 Marine Ecosystem

The behavior of bioplastics is less studied in marine environment than that of in other environments, and it is because of the fact that their fate is to be treated in soil or in different facilities of solid waste treatments. The important thing about biodegradable plastics is that they are not designed to degrade in the aquatic system and rather they are made to biodegrade in soil [29].

Degradation of bioplastics in the marine ecosystem is entirely different from the terrestrial environment. On the coastal line, the plastics convert into fragments rapidly because of their exposure to ultraviolet radiations whereas when they are submerged in sediments and dipped into water the process slows down. Biodegradation of bioplastics in the sea is incomplete because of lack of natural conditions and the nature of the polymer. For instance, the degradation rate of polyethylene (PE) is higher in tropical areas because of high temperature, suitable microorganisms and also high dissolved oxygen. Many living aquatic species are affected by marine plastic, mostly due to blockage of the gut. The plastic mortality rate is not well established, but it is considered as an important factor in many cases. A study shows that the biodegradable polymers such as shopping/carrier bags cause weight loss in 8.5% green turtles and 4.5% loggerhead turtles after 49 days [30].

3.4.2 Soil Ecosystem

According to various studies, it has been revealed that in the soil environment, many microorganisms degrade bioplastics [31–33]. The degradation rate of bioplastics in soil depends on the main components present in bioplastics and does not affect the circulation of nitrogen in soil except that of polylactic acid fiber (PLA). The soil with higher bacterial biomass and of powdered bioplastics has higher biodegradation rate in agricultural field [34]. Polylactic acid fibre (PLA), Polyhydroxybutyrate (PHB) and Polycaprolactams are nontoxic and have a degradation period of 2–3, 2–4, and 2–6 weeks respectively in soil. Degradable plastic polymers provide a gentle solution for waste management and pollution reduction [35]. In waste management, the fate of bioplastic in soil is depended upon its time of decomposition, and it may be aerobic or anaerobic decomposition. In aerobic decomposition, it is converted to carbon dioxide, compost and water while in anaerobic decomposition it is converted into carbon dioxide, methane and cell biomass preferably [19]. A biodegradable plastic seems to be more beneficial, but when they are disposed they decompose into air pollution.

4 Methods for Bioplastic Production

4.1 Production from waste materials

4.1.1 From Agriculture Waste

The agriculture waste can produce fibers and bioplastic polymers which are eco-friendly and biodegradable, and in the future bio-fibers would replace the petroleum-based polymers. Using this method, we can efficiently manufacture cellulose acetate fibers from cotton linters and flax fibers. Bioplastic based on cellulose acetate proved to be degraded within a year in both water and soil matrix. These materials can also be recycled and incinerated with leaving no ashes [36]. These studies also indicated that cellulose acetate has many significant chemical and physical properties such as colorability, strength, transparency, moldability, impact resistance, dielectric strength and versatility [37]. Natural bioplastic has many benefits and advantages which are manufactured in fluid form and can easily be molded without a high amount of pressure and energy source [38].

The procedure of making cellulose acetate is started with washing and bleaching the cotton linters and flax fibers with 120 mL bleaching agent which consists of 5.0% NaOH and 5.0% NaOCl to remove colors, fats and dust. In the next step after drying each sample of raw material, size is maintained to 35 g of raw material. After preparing the desired sample, another mixture of 10 mL prepared consisted of sulfuric acid and acetic anhydride with glacial acetic acid of 100 mL. The mixture is cooled at 7^0C, and raw material is added in mixture slowly with agitation to start acetylation process. This step leads to produce primary cellulose acetate which is in fluid form. The hydration of primary cellulose acetate is achieved by dilution with 30 mL of the equal part to 99.8% concentrated acetic acid and then rest of the mixture for 15 h to grow. Then, the resulting fluid was centrifuged in order to separate the final product. To the processed sample then added polyethylene glycol 600 as a plasticizer 25% by volume to cellulose acetate fluid and dried in the oven at 60 °C until the weight is constant to achieve the end product. For molding and shaping, product must be diluted with acetone which provides smooth and easy surfaces for pouring it. The production rate of the method 81% cellulose acetate is by flax fiber and 54% by cotton linters in viscous acetone-soluble fluid. In terms of biodegradable properties, these polymers lose 41–44% weight after 14 days [39].

4.1.2 From Vegetable Waste

Cereal and vegetable waste can also be used for producing bioplastic on the basis of their mechanical properties. The basic method accounts for the digestion of waste, such as cocoa pod husks, green herbs and rice hulls, in trifluoroacetic acid (TFA) [40]. The method starts with the fabrication and film casting process in which various types of the wastes were firstly dried at about 40 °C to remove any excess moisture and water content. Later, the material is added into a 60 mL vials, and a solution of 3% by weight is prepared with trifluoroacetic acid (TFA). The container is sealed with Parafilm, and the waste starts to dissolve in 3 days, but this time span can be different for different types of waste. The microcrystalline cellulose (MCC) in spinach and barley produces a viscous solution within a 3-day span. This solution is then centrifuged to remove residual material and then placed in a Petri dish for a few hours. This casting procedure is required to homogenize the solution, and all the films can be cast after 29 days of aging. The films are then kept in a vacuum hood so the solvent evaporates and then kept in 60% humidity for 2 days to get rid of TFA residue [41].

Chemical and morphological studies are important to find out whether the waste has been converted into acetylated cellulose. This was carried out by using Fourier transform infrared attenuated total reflectance (FTIR-ATR) spectroscopy. Another important parameter is finding the amount of water absorbed by the bioplastic. This process is carried out by putting the samples in humidity chambers for several days and comparing the dry weight with the final weight after water absorption. A Mettler Toledo thermogravimetric/differential scanning calorimetric (TGA/DSC) is used for finding the required thermo-mechanical characteristic of the bioplastic. This technology provides the decomposition rate of the bioplastic material in heating stress. The experiment proceeds with the formation of thermogravimetric curve used for weight loss and a derivative thermogravimetric curve used for the first derivative which is recorded with a specific time and temperature [42]. These bioplastics have a lot of advantages over the other bioplastics. They have a varied form of rigidity and fragility to being stretchable and soft at the same time. Natural amorphous cellulose is formed when we mix the vegetable waste solution in a calculated amount of TFA which is a property we need to conserve the environment from harmful plastics [43].

4.1.3   From Banana Peels

In the given method, banana peels, HCL, sodium metabisulfite, glycerol, distilled water and starch materials were used to synthesize bioplastic [44]. First of all, banana peels were converted into small pieces and it was soaked in a 0.2 M solution of sodium metabisulfite for 45 min. In distilled water, banana peels were boiled for about half an hour. Peels were left for half an hour on a filter paper until dry. Then, a uniform paste of banana peels was formed by blending in a beaker. A 25 g of banana paste was placed in a beaker with 6 mL of HCl, glycerol (2 M) and NaOH (0.5 N) was added for adjusting pH. Finally, the whole material was poured in a Petri dish, placed in an oven for half an hour at 130 °C. The tensile strength increases with the presence of starch which was measured by using standard machine ASTM [30]. It was also observed that the degradation process starts after 9 months because of microorganisms' growth and the total period observed for biodegradation was from 7 to 9 months. Malleability and resilience are one of the major advantages of making plastic tubes and bags along with its biodegradation in the environment without producing any pollution. Potato peels can be used in the future because it contains a high amount of starch and polymer chains with a maximum 90% efficiency as compared to banana peel efficiency which is 80%. The main reason for no usage is because potato peels take more time to be dried than banana peels which require only one day. These bioplastics can be used as a bag or for packaging, as the use of sodium metabisulfite and glycerol is to prevent the production of microorganisms and to increase the malleability, respectively [45]. The main problem of bioplastic production through banana peels is to maintain the accordance of biodegradation in spite of its importance [46].

4.1.4 From Melt Recycling of Plastic Waste

Well ground polylactic acid in the form of flakes was used shifting into the extruder hopper. After, it dried the polylactic acid flakes in the oven at 82 °C for 14 h before melting and spinning process. In this technique, the normal water fixation in the wake of drying was 100 ppm [47]. Before the melt spinning, crystallization and drying strategies were completed at 160 °C (for 4 h).

The recycled and dried PLA flakes were melt spun by using a laboratory single screw extruder with a screw diameter of 6.35 mm and a 1-mm die hole diameter. The extruder was outfitted with a cooling coat to keep the reused flakes away from melting inside the feed container [48]. Two heating bands were situated along the screw and one on the die. The warming temperatures were kept at comparative esteem. The produced fiber was first spun through a single-nozzle circular (1 mm breadth) and afterward gathered on a hardened steel drum situated at 53 cm with a speed of 70 m/min. Diverse temperatures from 180 to 240 °C were balanced for assessing the impacts of preparing condition on the melt spinning of the polylactic acid strands. After the melt spinning of polylactic acid, fibers were drawn with a mechanical testing machine with the consistent extension rate of 50 cm/min [47].

The polylactic acid fibers were colored by the exhaust dyeing technique. The liquid proportion was acclimated to 1:50. The dying bath contained 0.5 g/L Avolan IS as dispersant operator, and the bath acidity was 5.0 (formic acid). The dyeing temperatures for polylactic acid were 110°C, separately. The temperature for polylactic acid dyeing was set as prescribed by Lunt and Bone [49]. After that, fibers from the dyeing bath was removed and flushed with water and treated in a soaping shower 1.5 g/L nonionic surfactant utilizing liquid proportion of 30:1 at temperature of 40 °C for 30 min. At that point, they were washed with soft water and dried in normal conditions [50]. After that, biodegradable fibers are ready to use and they find their applications in medical textiles, biomedical implants and drug release systems.

4.2 Production from Wastewater

4.2.1 From Activated Sludge

This method involves the production of polyhydroxyalkanoates (PHAs), which is a biodegradable plastic through the use of activated sludge from wastewater treatment plants of variable nature such as starch (SAS), municipal (MAS), dairy (DAS) and pulp–paper (PAS) [51]. Wastewater is used as a carbon source where gravity settling chamber is used to measure the concentration of suspended solids present in activated sludge [52]. If wastewater is used as carbon source, then known volume of sludge is added to wastewater on order to get 0.5 g/L concentration of suspended solids. The pH 7.0 is adjusted, and the flasks in a rotary shaker are incubated for 24 h at 25 °C. The samples are taken out from the incubation chamber at a predetermined time [53].

It has been found that paper industry effluent has the highest tendency to form PHA among all the four types of wastewaters described in this article. It is attributed to the extensive presence of volatile fatty acid, known as a substrate which is essential for the production of PHA [54].

Scientific investigations report that industrial wastewater has a high concentration of volatile fatty acids so it is considered as a good source for the production of the biopolymer. Therefore, this method offers dual benefits, i.e., production of bioplastic along with the treatment of wastewater. Following this approach, 40% removal of COD could be achieved with consequent formation of PHA which reduces the pollution load and treatment cost of the treatment plant [55].

4.2.2 From Dairy Wastewater

Production of bioplastics or polyhydroxyalkanoates (PHAs) by using milk whey and activated sludge from dairy wastewater is another potential approach. Microorganisms present in activated sludge use lactose (milk sugar) as a nutrition source [41]. When the pre-treated milk whey from thermal treatment is added to the saline medium in the C:N ratio of 50%, fermentation starts and PHAs are produced readily by microorganisms. After stopping the fermentation process, the concentration of lactose is reduced in an optimized pH medium and the production of bioplastics slows down, while the medium without pH adjustment has a lactose concentration of 6.17 g/L and the production of bioplastics increases to 13.82%. It is clear that the maintained pH medium is not suitable for this process, since it hinders the efficient production of PHA [56] (Table 1).

Table 1 Effect of pH adjustment on bioplastic production [56]

Sample	% PHAs	PHAs (g/l)
K-milk (pH adjusted)	5.10	0.072
K-milk (no pH unadjusted)	13.82	0.224

4.2.3 From Cassava Starch Wastewater

In this method, a sequencing batch reactor (SBR) treatment system is used for making polyhydroxyalkanoate (PHA) in which the primary ingredient used was cassava starch wastewater. The system was seeded with a PHA producing bacterial strain, *Bacillus tequilensis* MSU 112. The manufacturing process with initially inoculums required some preparation process. Cassava pulp is taken from cassava starch industry for the isolation of *Bacillus equilenin*, and a culture medium is produced. The medium is mainly composed of 2.5 g/L glucose and dipotassium hydrogen phosphate, 5.0 g/L sodium chloride, 17.0 g/L tryptone and 3.0 g/L phytoene peptic digest of soya meal. The medium is expanded at 30 °C for a day with an agitation rate of 200 rpm, and then the cells are centrifuged and used as inoculum [57].

Cassava starch was added in different amounts of 3.0–5.0 g/L with its pH adjusted between 7 and neutral. The 3 g/L CSW is added in the reactor along with 2% inoculum culture. After this, the reactor was operated batchwise along with constant stirring for mixing up material to obtain activated sludge. Proceeding with this process for days, the reactor is then operated according to sequenced batch with filling face for 20 min, reaction phase for 22 h, settling phase 1 h and withdrawal phase 40 min completing in a 24-h cycle. Anoxic/aerobic steps of 4/18 h were used during the reaction phase at all COD concentrations maintaining the dissolved oxygen concentration at 1–2 mg/L. All the cycles are proceeded with the usage of 1.0 L volume of activated sludge and 4.0 L fresh CSW. When the sequence batch reactor (SBR) system proceeds for 7 cycles, steady state is achieved. The reactor still functioned for 4 and 5 g/L concentration of COD, while the mixed liquid suspended solid was maintained between 1.5 and 5.0 g/L. SBR treatment system in the study provides for a good functioning of CSW usage in the production of PHA. Furthermore, seeding of *Bacillus tequilensis* MSU 112 increases the production of PHA for more plastic production [58].

4.3 Microbial Synthesis of Bioplastics

4.3.1 From Engineered Bacteria

Scientists used recombinant for the production of PHA, because of its suitability for fast growth, genetic manipulation, great cell density and ability to consume carbon. For the production of recombinant *Escherichia coli,* before this, it requires PHA synthetic gene which is Polyhydroxybutyrate synthesis genes (phbCAB) from *R.*

eutropha bacteria. The hemoglobin gene vgb of bacteria manufactures polyhydroxybutyrate (PHB) with dry cell weight of 206 g L^{-1} which holds 73% PHB. This process productivity rate is 3.4 g L^{-1} per hour [59]. In starch agar, recombinant L is capable to produce 167 g L^{-1} PHB with a productivity rate of 3.05 g L^{-1} per hour and additional process which is temperature inducing treatment can recover 95% PHB from agar solution [60].

Recombinant *E. coli* can be very productive to manufacture other polymers like (R)-3-hydroxyhexanoate (HHx), copolymer P3HB4HB, poly-4-hydroxybutyrate homopolymer, (R)-3-hydroxyvalerate (HV) and (R)-3-hydroxybutyrate [61]. If the beta oxidation gene fadB was deleted in the genome of bacteria, this condition becomes very useful for recombinant *E. coli*. Because of the phaC1 gene of recombinant, *E. coli* was adept to manufacturing medium length sequence PHA from fatty acid [62].

4.3.2 Polyhydroxyalkanoates (PHAs) from Microbes

PHA is completely biodegradable and versatile polymers [63]. Microbes are the major source for the production of bioplastics. They are biologically synthesized by a variety of gram-negative and gram-positive bacterium which may include *Bacillus* sp., *Pseudomonas* sp. and *Methylobacterium* sp. [64, 65]. Microbes perform the metabolic activity, biologically synthesize PHAs and store them in the cytoplasm as small granules in the form of carbon or energy compounds [66, 67]. Mostly, they are produced under nutrient-limited conditions like when oxygen, nitrogen or phosphorus deficiency prevails in their bodies while carbon is excessively available. However, some bacteria like *A. eutrophus* and *A. latus* can also produce PHAs under un-stressed conditions [68].

The choice of microbe is a major factor; i.e., we have to choose the bacterium which has faster growth rate so that it carries out metabolic activity at a faster rate which will lead toward rapid production of bioplastics. The second factor is the choice of media which must be selected carefully [69]. Media must be cheap and able to produce the product in amounts that is economically beneficial [70]. Extensive studies are being carried out to select the cheap sources for PHA production which shows that activated sludge [52], starch wastewater [71], molasses [72], milk whey, rice and wheat bran (wastewater from palm and olive oil mills) and swine waste is the most common and cheap fermentation media [73, 74].

Fermentation is carried out at an optimum temperature usually at 37 to 38 °C. The pH may be controlled or may leave uncontrolled. Low stirring is carried out to keep the dissolved oxygen concentration at a lower level. For PHA production batch, fed-batch and continuous fermentation [75] methods are considered to be more efficient. Recovery is the last step which must be carried out in a way that is safe, cheap and economical for industrial production of PHAs. Organic solvent extraction is the most commonly used method. This method is far better than cell disruption because in this molar weight of biopolymer decreases [76]. PHAs found their applications in many fields. Earlier, they are used in packaging like to prepare shampoo bottles, containers for cosmetics, covers for paper and cardboard, milk cartons, pens, combs, bullets, etc. Recently, PHA found their applications in the medical field like used for the production of cardiovascular products [77].

4.3.3 PHA Production by Hydrocarbon Degraders

Polyhydroxyalkanoate (PHA) production can also be affected by fluctuations in the environment, where a certain type of xenobiotic substances compels the organisms, increasing the PHA production [78]. A more precise example can be that sites containing oily pollutants have limited nitrogenous compounds (less than 1%) and more carbon content (84%), which are suitable in increasing the cell production for PHA [79]. The PHA is produced by a vast variety of bacterial strains, while oil is being degraded. These include Brochothrix, Caulobacter, Ralstonia, Yokenella, Acinetobacter, Burkholderia and Sphingobacterium [80].

Ralstonia eutropha JMP 134 produces PHB when growth-inhibiting substances like sodium benzoate are provided as a source of carbon under a condition of nutrient stress or limitation. The described experiment has given attention to reduce toxicity and use microbes for valuable bioproduction [46]. Another substance by the name of Bacillus cereus FA11 has also been reported to produce a copolymer. It is isolated from polluted soil containing trinitrotoluene which is heated at the temperature of 30 oC and at the pH 7 producing 48.43% of the copolymer (3HB-co-3HV) [81]. A marine bacterium named as Alcanivorax borkumensis SK2 has also been discovered to produce PHB depositions on aliphatic hydrocarbons. When toluene is presented as a carbon source, a bacterium *Rhodococcus aetherivorans* IAR1 instincts to produce copolymer along with triacylglycerol. This functions as a cost-efficient and environmental-friendly procedure to remove a certain type of waste [82]. All of these procedures and many more are extremely helpful in producing highly valued exopolymer and end polymer by using the same organism simultaneously while adjusting the optimum temperature and other environmental aspects [83, 84].

4.3.4 Photosynthetic Bacteria as PHA Producers

Cyanobacteria are the photosynthetic prokaryotes microorganisms having the ability to produce and restore PHA in bodies by using oxygenic photosynthesis process. Most cyanobacteria are known to produce polyhydroxybutyrate in deficient conditions of phosphorus [85]. The species *of Nostoc muscorum, Synechococcus and Spirulina platensis* are mostly known to generate polyhydroxybutyrate under limited stress condition of phosphorus [86]. Sometimes, the limited nitrogen condition is also helpful to generate polyhydroxybutyrate in *Synechocystis* sp. PCC 6803 and *Synechocystis* sp. UNIWG, up to 15% and 14%, respectively [87]. The *Synechocystis* sp. PCC 6803 also increased the production process of polyhydroxybutyrate if there are limited conditions in gas exchange, nitrogen and phosphorus [88]. The studies reported that the depressed condition of sulfur enhances the polyhydroxybutyrate yield by 3 and halftimes [89]. A study described that in high chemoheterotrophy, mixotrophy conditions *N. muscorum* could produce five times more polyhydroxybutyrate with limiting nitrogen state [90]. The other major important factors which play a role in the production of polyhydroxybutyrate in *N. muscorum* are sunlight, pH, phosphorus, light–dark cycles, nitrogen, and exogenic carbon status and sources [91].

4.3.5   PHA Production from Halophiles

Archaeon is a bacterium which lives in marshlands, hot water springs, salt lakes and oceans. They are also known as extremophiles because they are adapted to live in extreme environments. This bacterium has the ability to produce PHA a source of bioplastic. They require salty conditions for their growth [92]. PHB is produced by extremely halophilic archaebacteria (Halobacteriaceae) under stressed conditions, i.e., in the limited availability of nutrients and excess availability of carbon [93]. Another bacterium named as Haloferax mediterranei which grows best at 25% concentration of salt is able to produce 60–65% of PHA when grown in a limited supply of phosphate [94]. Moderately, halophile bacterium Halomonas boliviensis LC1 which grows at 3–15% concentration of salt leads to the production of PHB up to 56% by utilizing starch hydrolysate as a source of food. Bacterium extracts maltose from starch. If the supply of oxygen is limited, then it will enhance production. If sodium acetate and butyric acid are used as a carbon source and the nutrient supply is limited then H. boliviensis LC1 leads to the production of large amount of PHB up to 88% of cell dry weight during the stationary phase [95]. One of the bacterium *Haloarcula marismortui* which is extremely halophilic in nature is able to produce PHA by using vinasses which is the by-product produced in ethanol manufacturing industries. This method involves the use of shake flask experiment. Use of raw vinasses up to 10% leads to the production of almost 24% of PHB [96].

4.3.6   Plant Growth Promoting Rhizobia (PGPR) as PHA Producers

The term rhizosphere is the soil neighboring to the plant's root which may harbor organisms upgrading the development of roots and plants by discharging extracellular metabolites [97]. The nutrients discharged by the root exudate to its surrounding rhizosphere; then in turns, it serves as a hotspot for microbial interaction and thus nutrient recycling happens [98]. It is plainly explained that rhizosphere has hidden reservoirs for polyhydroxyalkanoate accumulators in addition to the plant growth promoting rhizobia and hostile impacts. Some of the organisms, such as *Burkholderia terricola, Lysobacter gummosus, Pseudomonas extremaustralis, Pseudomonas brassicacearum* and *Pseudomonas orientalis*, have been accounted for as polyhydroxyalkanoate makers dependent on polymerase chain reaction method having polyhydroxyalkanoates as the targeted gene [99].

Production of the polyhydroxyalkanoates holds invaluable attributes toward upgraded root colonization, plant development advancement, survivability, chemotaxis, motility and cell duplication [100]. Non-rhizosphere soil has low polyhydroxyalkanoate production than rhizosphere soil. This production is based on cultivation-dependent techniques. Independent cultivation techniques enabled us to reason that wheat, oilseed assault and sugar beet rhizosphere have a more polyhydroxyalkanoate generation [101, 102]

4.4 Plant-Based Bioplastics

4.4.1 Polylactic Acid Production from Plants

Polylactic acid is created by normally happening lactic acid bacteria of the family Lactobacillus, which mature hexose sugars. Lactate dehydrogenase changes over pyruvate into lactate, with corresponding oxidation of nicotinamide adenine dinucleotide hydrogen [103]. The other technique utilized by Dow Cargill to create their Nature Works PLA item includes the condensation of lactic corrosive to lactide, pursued by ring-opening polymerization to deliver high-atomic mass polylactic acid [104].

Ongoing development in polylactic acid creation has been made by using a 'microbial manufacturing plant' rather than ring-opening polymerization for the fermentation of high-sub-atomic mass polylactic acid. A Japanese group has changed a polyhydroxyalkanoate synthesizing catalyst to encourage polymerization of lactic acid, consequently permitting polymer synthesizing inside the microorganism without the requirement for compound polymerization [105, 106].

4.4.2 Soy-Based Bioplastics

There are some steps which are required for the synthesis of bioplastic from soybean in which mostly the protein content of soybean is used. The first step is the breaking of intermolecular bonds with low energy which keeps the polymers stable that are broken down in the first step. Then in the next step, arrangement and orientation of polymer chains occur. In a third step, a three-dimensional network is formed due to new bonds and interactions which occur after removal of the agent which break down the intermolecular bonds [107].

Films are made by two important processes. The first step is a wet process in which solubilization and dispersion of proteins occur. Material formation from protein dispersion has widely studied [18]. In this process, a thin layer of protein solution is formed. PH is also a significant factor. For unfolding of protein, generally dispersion is occurred in alkaline PH. Protein sensitivity to PH change is related to great content of ionized polar amino acids. If great content of ionized polar amino acids is present in soy proteins, then film formation will be limited at low PH [108].

Next is a dry process which is dependent on the thermoplastic properties of proteins under less water content. Thermo-mechanical and thermal processes are used to study the thermoplastic behavior of proteins for film formation under low water content conditions. According to glass transition theory, changes in texture occur during thermoplastic polymer processing. In this, specific glass transition temperature is given which changes glassy state to rubbery state. So at glass transition temperature, when protein-based polymers are heated a soft material is obtained which can be used in packaging materials [109, 110].

4.5 From Natural Polymers

4.5.1 Wheat Gluten

Wheat gluten is obtained from the bioethanol industry. It is available abundantly, at low cost, and mostly it is used for baking industry and animal feed. It has been used because of its nature for the formation of films, and its mechanical properties are suitable for applications like the packaging. Its structural conformation is highly important during the production of plastic and baking. However, its structure is affected by many different factors during growth of the plant. During the production of plastic, changes in its structure are done by plasticizers, chemicals and temperature. At high temperature, it also has shown positive effects on materials' properties, as it converts from unorganized to organized form [111].

4.5.2 Potato Starch

Starch is a biodegradable polymer which is stored in plants and used for short-time applications, for instance, packaging. For better tensile and gas barrier properties, starch granules break down to a thermoplastic material by heating, mixing and stress. Oxidized potato starch along with starch with high amylose has been used for plastics as raw material to improve the tensile properties, and they are much better than the original ones. Improved materials' properties are because of the high content of amylose in starch or amylopectin fractions with changed chain lengths. The high temperature is also required for higher material performance [71, 112, 113].

4.6 Bioplastic from Starch

In this, thermoplastic was made using starch which is easily available in plants. Polylactic acid, cellulose acetate and viscose are some semisynthetic bioplastics. They are commercially significant because of their strength in many packaging applications. To increase the biodegradability and to reduce cost, it is combined with starch but

PLA is still very expensive. For this purpose, starch is used for the short period used applications such as packaging as a raw material.

For the preparation of bioplastic from starch, first of all, cornstarch before use was dried at 70 °C for 24 h in an oven. The chemicals that were used are urea, propylene glycol, glycerol, choline chloride and ethylene glycol. Four deep eutectic solvents (DES) that were used are ChCl: 2 propylene glycol, ChCl: 2 ethylene glycol, ChCl: 2 urea and ChCl: 2 glycerol in 1:2 molar ratio of choline chloride and hydrogen bond. Both DES and corn starch were mixed using a food processor of Kenwood BL330 Series, and then it was placed for 3 h in an oven at 50 °C and immediately poured in a bag. By using the same mixing procedure of DES and baking starch, three components Glycerin 200, filler and starch were mixed by the difference of cornstarch amount with filler as it depends on the experiments. Fillers that were used are orange and banana peels, eggshells, wood–fiber, wood–flour and silk powder. Lignin was mixed with Glycerine 200 for the formation of lignin adhesive in a beaker and placed in an oven for 24 h at 50 °C after mixing it manually.

For mechanical analysis, samples were prepared using a compression molder by making TPS sheets. The mixture was placed between two copper plates with a copper separator of a 1 mm, and then it was placed with a force of 120kN hydraulic press at various temperatures for 10 min. Water was passed to cool it down by maintaining the pressing force. Thermoplastic starch containing cornstarch and thermoplastic based on polyolefin are comparable. TPS used for different applications has mechanical properties which are acceptable when used for a short period of time [114].

4.7 Bioplastics from Cellulose Diacetate

Cellulose diacetate (CDA) was perceived as a biodegradable polymer, and different preliminaries have been undertaken to give adequate thermoplasticity to cellulose acetates with the end goal to render them dissolve in any process. This is on the grounds that cellulose diacetate, which has the best thermal plasticity among a wide range of cellulose acetates, neglects to indicate satisfactory dissolving conduct without disintegration or coloring. Thus, lessening the flow temperature of cellulose acetates is vital and it requires the expansion of plasticizers and flows promotors. At present, phthalates and phosphates are utilized in modern way by different procedures which are very time taken (i.e., 4–5 h for each batch). These plasticizers are not appropriate for the arrangement of biodegradable polymers due to the unsafe natures of their disintegration items. In this association, there have been a few attempting to use aliphatic polyesters of bacterial causes and in addition synthetic ones as plasticizers for cellulose acetate [115].

Now authors try to form the methodology concerning with the plasticization of cellulose acetates during the melting processing by reaction with dibasic acid anhydrides and monoepoxide. For this situation, oligoesterified cellulose acetates, produced by the reaction of cellulose acetates with succinic anhydride and phenyl glycidyl ether, for instance, at a temperature somewhere in the range of 70 and

180 °C, have the theoretical structure. This sort of introduction of the oligoester side chain into the cellulose acetate atom would improve the thermoplasticity of cellulose acetates. While when it was responded with succinic anhydride and phenyl glycidyl ether at 120 °C for 20 min under plying conditions, it was changed over easily into a thermoplastic material. Accordingly, the above supposition that the arrangement of oligoester chains as joined parts of together with homo-oligomers upgrades the thermoplasticity of the items was affirmed. Accordingly, the above supposition that the arrangement of oligoester chains as joined parts of cellulose acetates together with homo-oligomers upgrades the thermoplasticity of the items was confirmed [116].

5 Biodegradable Plastic and Green Economy

5.1 Environmental Benefits

Due to a dramatic increase in population with the advancements in their lifestyle, needs and demands of humans are also been increased. For the fullness of their needs, a large number of industries are under operation. As a result of this, accumulation and the production of waste are also been increased. From these industries, plastic industries are the most important and widely used industry. Because of this, many environmental problems originate. But from recent decades, researchers and scientists are hardly trying to find the solutions to sustain the environment. Now they start trying to move toward the green environment. They find the new techniques and methods for the formation of plastic from renewable resources which are environmentally friendly.

They use a large number of microorganisms which produced naturally polyhydroxyalkanoate (PHA) and polyhydroxybutyrate (PHB) plastics. These plastics are environmentally friendly and used a lot in our daily lives. These plastics are completely biodegraded in one year by the use of microorganisms and turn to carbon dioxide and water and then to nature [117].

Bioplastics are playing a major role in the reduction and in the treatment of waste. A lot of materials which we used in our daily lives are made from biodegradable plastics such as packaging material, fibers and bags. From the usage of these materials, waste minimization occurs and it also has been reducing the burden of waste from the environment.

5.2 Green and Sustainable Methods to Generate Plastic

From the last few decades, bioplastics take the place of those plastics which are made from fossil fuels and petroleum. Scientists and researchers used great sustainable methods for the generation of bioplastics. They used different microorganisms such as *Rhizobacteria, Burkholderia terricola, Lysobacter gummosus, Pseudomonas extremaustralis*, and plants such as family *Lactobacillus* to make the biodegradable plastics. These plastics are degraded in one year and are environment-friendly [99, 103].

5.3 Renewable Resources Are Used

Microorganisms are equipped for creating polyhydroxyalkanoates from different carbon sources extending from cheap, complex waste effluents to plant oils, unsaturated fats, alkanes and in addition basic sugars [118–120]. Every year, a lot of waste materials are released from rural and nourishment preparing businesses and these wastes speak to a potentially sustainable feedstock for polyhydroxyalkanoate creation. Using these waste materials as carbon hotspot for PHA creation decreases the substrate cost, as well as recovers the expense of discharge of waste [121, 122].

5.4 Social Benefits

Bioplastics play a major role in society because of its applications in all areas of daily life such as, from technical applications like automobiles and medical to packaging material for food. Through bio-economy as a driver for a new job, they provide benefits for different stakeholders such as workers. It brings a positive change in the behavior of society [123].

5.5 Economic Benefits

Bioplastics are also economically beneficial. It is used in various applications which are economically profitable. It provides a green economy. We get different materials from it such as fibers, bags, shampoo bottles, medical or surgical instruments, home instruments and fertilizer bags. In one way, bioplastics are cost-effective if we see them as environment-friendly. They save our money in that way, by reducing the cost which we serve on climatic problems. But in other ways, it is more expensive than standard plastics [124, 125].

6 Comparative Analysis of Plastics and Bioplastics

See Fig. 3.

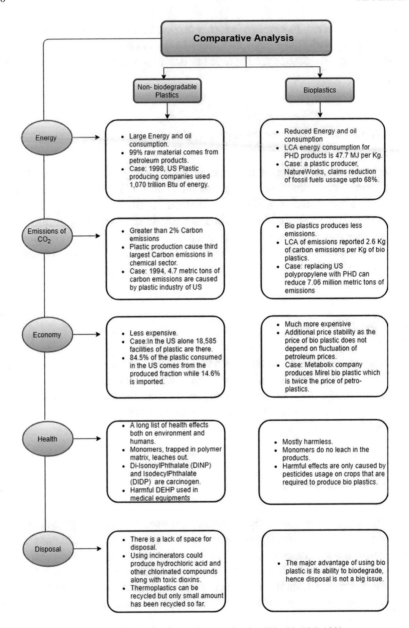

Fig. 3 Comparative analysis of bioplastics and petro-plastics [22, 64, 126–139]

Limitations and Obstacles in the Use of Bioplastics
They are many problems and limitations in the use of bioplastics, and some of them are given below:

- Bioplastics are costly as compared to common plastic.
- Bioplastics can spoil the recycling process if they are not separated from common plastics.
- Bioplastic production from renewable resources can reduce the reserves of raw materials.
- Bioplastic description can be confusing as people might relate it to composting, and some manufacturers use it as a tactic to lure public for their products as non-toxic, eco-friendly, etc [140].
- Bioplastic is not a solution of marine littering as people still throw their waste mainly plastic (bags, straws, etc.) on seashores.
- Bioplastic usage encourages people to more littering instead of less use of any kind of plastic as it pollutes the environment whether it takes less or more time to degrade.
- Bioplastic land requirement competes with the land producing food as the crops can be used as a food resource for people instead of using them for bioplastics [30, 141, 142].

7 Conclusion

There are numbers of methods and technologies for the synthesis and production and of bioplastics. If people have more awareness of environmental issues, then society can be more conscious about the environment and its sustainability. Plastic is the basic and prime raw material in every industry and even in the domestic living of community. But nobody knows those problems which are caused by non-biodegradable plastic. Only awareness is required for society to invest in bioplastic and change the structure of the economy. There are many benefits of bioplastic such as manufactured from renewable resources, biodegradable, as well as able to be recycled, burned and composted without generating any harmful by-product in the environment. The bioplastic can reduce carbon emissions and footprints to produce low carbon finger printing on environment. Bioplastic may not be one-stop solution of all environmental-related problems, but it is the fastest way that leads to the sustainable development of society coping the environmental issues and problems.

Acknowledgements The authors are highly thankful to the University of Gujrat, Gujrat, Pakistan.

Conflicts of Interest The authors declare no conflict of interest.

References

1. Rosalia Alvarez-Chavez, C., Edwards, S., Geiser, K., Rosalía Álvarez-Chávez, C., Moure-Eraso, R., Geiser, K.: Sustainability of bio-based plastics: general comparative analysis and recommendations for improvement Author's personal copy sustainability of bio-based plastics: general comparative analysis and recommendations for improvement. J. Cleaner Prod. (2011). https://doi.org/10.1016/j.jclepro.2011.10.003
2. Halden, R.U.: Plastics and health risks. Annu. Rev. Public Health **31**, 179–194 (2010). https://doi.org/10.1146/annurev.publhealth.012809.103714
3. Philp, J.C., Ritchie, R.J., Guy, K.: Biobased plastics in a bioeconomy. Trends Biotechnol. **31**, 65–67 (2013)
4. Law, K.L., Morét-Ferguson, S., Maximenko, N.A., Proskurowski, G., Peacock, E.E., Hafner, J., Reddy, C.M.: Plastic accumulation in the North Atlantic subtropical gyre. Science **329**, 1185–1188 (2010). https://doi.org/10.1126/science.1192321
5. El-Kadi, S.: Bioplastic production from inexpensive sources. In: Bacterial Biosynthesis, Cultivation System, Production and Biodegradability, p. 145 (2010). ISBN 9783639263725
6. Peelman, N., Ragaert, P., De Meulenaer, B., Adons, D., Peeters, R., Cardon, L., Van Impe, F., Devlieghere, F.: Application of bioplastics for food packaging. Trends Food Sci. Technol. **32**, 128–141 (2013)
7. Mohanty, A.K., Misra, M., Drzal, L.T.: Sustainable bio-composites from renewable resources: opportunities and challenges in the green materials world. J. Polym. Environ. **10**, 19–26 (2002). https://doi.org/10.1023/A:1021013921916
8. Reddy, R.L., Reddy, V.S., Gupta, G.A.: Study of bio-plastics as green & sustainable alternative to plastics. Int. J. Emerg. Technol. Adv. Eng. **3**, 82–89 (2013)
9. de Koning, G.J.M., Lemstra, P.J.: Crystallization phenomena in bacterial poly[(R)-3-hydroxybutyrate]: 2 embrittlement and rejuvenation. Polymer **34**, 4089–4094 (1993). https://doi.org/10.1016/0032-3861(93)90671-V
10. Flieger, M., Kantorová, M., Prell, A., Řezanka, T., Votruba, J.: Biodegradable plastics from renewable sources. Folia Microbiol. **48**, 27–44 (2003)
11. Gross, R.A., Kalra, B.: Biodegradable polymers for the environment. Science **297**, 803–807 (2002)
12. Bordes, P., Pollet, E., Avérous, L.: Nano-biocomposites: Biodegradable polyester/nanoclay systems. Progress in Polymer Science (Oxford) **34**, 125–155 (2009)
13. Shaikh, H.M., Pandare, K.V., Nair, G., Varma, A.J.: Utilization of sugarcane bagasse cellulose for producing cellulose acetates: novel use of residual hemicellulose as plasticizer. Carbohyd. Polym. **76**, 23–29 (2009). https://doi.org/10.1016/j.carbpol.2008.09.014
14. Suyatma, N.E., Tighzert, L., Copinet, A., Coma, V.: Effects of hydrophilic plasticizers on mechanical, thermal, and surface properties of chitosan films. J. Agric. Food Chem. **53**, 3950–3957 (2005). https://doi.org/10.1021/jf048790+
15. Bertan, L.C., Tanada-Palmu, P.S., Siani, A.C., Grosso, C.R.F.: Effect of fatty acids and "Brazilian elemi" on composite films based on gelatin. Food Hydrocolloids **19**, 73–82 (2005). https://doi.org/10.1016/j.foodhyd.2004.04.017
16. Sothornvit, R., Krochta, J.M.: Plasticizer effect on mechanical properties of β-lactoglobulin films. J. Food Eng. **50**, 149–155 (2001). https://doi.org/10.1016/S0260-8774(00)00237-5
17. Vieira, M.G.A., Da Silva, M.A., Dos Santos, L.O., Beppu, M.M.: Natural-based plasticizers and biopolymer films: a review. Eur. Polymer J. **47**, 254–263 (2011)
18. Donhowe, G., Fennema, O.: Edible films and coatings: characteristics, formation, definitions, and testing methods. In: Edible Coatings and Films to Improve Food Quality, pp. 1–24 (1994)
19. Šprajcar, M., Horvat, P., Andrej, K.: Biopolymers and bioplastics plastics aligned with nature. National Institute of Chemistry, Ljubljana, p. 32 (2013)
20. White, G.F., Russell, N.J.: What is biodegradation? In: Karsa DR, Porter MR (eds.) Biodegradability of Surfactants, pp. 28–64. Springer, Netherlands (1995)
21. Fritsche, W., Hofrichter, M.: Aerobic degradation by microorganisms. In: Biotechnology: Second, Completely Revised Edition, vol. 11–12, pp. 144–167 (2008). ISBN 9783527620999

22. Mohee, R., Unmar, G.D., Mudhoo, A., Khadoo, P.: Biodegradability of biodegradable/degradable plastic materials under aerobic and anaerobic conditions. Waste Manage. **28**, 1624–1629 (2008). https://doi.org/10.1016/j.wasman.2007.07.003

23. Ghuttora, N.: Increase the usage of biopolymers and biodegradable polymers for sustainable environment (2016)

24. Pathak, S., Sneha, C.L.R., Mathew, B.B.: Bioplastics: Its Timeline Based Scenario & Challenges. J. Polymer Biopolymer Phys. Chem. **2**, 84–90 (2014). https://doi.org/10.12691/jpbpc-2-4-5

25. Andrady, A.L.: Microplastics in the marine environment. Mar. Pollut. Bull. **62**, 1596–1605 (2011)

26. Pham, C.K., Ramirez-Llodra, E. , Alt, C.H.S., Amaro, T., Bergmann, M., Canals, M., Company, J.B., Davies, J., Duineveld, G., Galgani, F., et al.: Marine litter distribution and density in European seas, from the shelves to deep basins. PLoS ONE **9** (2014). https://doi.org/10.1371/journal.pone.0095839.

27. Sudesh, K., Iwata, T.: Sustainability of biobased and biodegradable plastics. Clean: Soil, Air, Water **36**, 433–442 (2008). https://doi.org/10.1002/clen.200700183

28. Emadian, S.M., Onay, T.T., Demirel, B.: Biodegradation of bioplastics in natural environments. Waste Manage. **59**, 526–536 (2017)

29. Tosin, M., Weber, M. , Siotto, M., Lott, C., Innocenti, F.D.: Laboratory test methods to determine the degradation of plastics in marine environmental conditions. Front. Microbiol. **3** (2012). https://doi.org/10.3389/fmicb.2012.00225

30. Chinaglia, S., Tosin, M., Degli-Innocenti, F.: Biodegradation rate of biodegradable plastics at molecular level. Polym. Degrad. Stab. **147**, 237–244 (2018). https://doi.org/10.1016/j.polymdegradstab.2017.12.011

31. Suyama, T., Tokiwa, Y., Ouichanpagdee, P., Kanagawa, T., Kamagata, Y.: Phylogenetic affiliation of soil bacteria that degrade aliphatic polyesters available commercially as biodegradable plastics. Appl. Environ. Microbiol. **64** (2018)

32. Teeraphatpornchai, T., Nakajima-Kambe, T., Shigeno-Akutsu, Y., Nakayama, M., Nomura, N., Nakahara, T., Uchiyama, H.: Isolation and characterization of a bacterium that degrades various polyester-based biodegradable plastics. Biotech. Lett. **25**, 23–28 (2003). https://doi.org/10.1023/A:1021713711160

33. Kim, M.N., Kim, W.G., Weon, H.Y., Lee, S.H.: Poly(L-lactide)-degrading activity of a newly isolated bacterium. J. Appl. Polym. Sci. **109**, 234–239 (2008). https://doi.org/10.1002/app.26658

34. Adhikari, D., Mukai, M., Kubota, K., Kai, T., Kaneko, N., Araki, K.S., Kubo, M.: Degradation of bioplastics in soil and their degradation effects on environmental microorganisms. J. Agric. Chem. Environ. **05**, 23–34 (2016). https://doi.org/10.4236/jacen.2016.51003

35. O'Brine, T., Thompson, R.C.: Degradation of plastic carrier bags in the marine environment. Mar. Pollut. Bull. **60**, 2279–2283 (2010). https://doi.org/10.1016/j.marpolbul.2010.08.005

36. Ach, A.: Biodegradable plastics based on cellulose acetate. J. Macromol. Sci. Part A **30**, 733–740 (1993). https://doi.org/10.1080/10601329308021259

37. Yuan, J., Dunn, D., Clipse, N.M., Newton, R.J.: Formulation effects on the thermomechanical properties and permeability of free films and coating films: characterization of cellulose acetate films. Pharm. Technol. **33**, 88–100 (2009)

38. Qiu, X., Hu, S.: "Smart" Materials based on cellulose: a review of the preparations, properties, and applications. Materials **6**, 738–781 (2013). https://doi.org/10.3390/ma6030738

39. Mostafa, N.A., Farag, A.A., Abo-dief, H.M., Tayeb, A.M.: Production of biodegradable plastic from agricultural wastes. Arab. J. Chem. **11**, 546–553 (2018). https://doi.org/10.1016/j.arabjc.2015.04.008

40. Bayer, I.S., Guzman-Puyol, S., Heredia-Guerrero, J.A., Ceseracciu, L., Pignatelli, F., Ruffilli, R., Cingolani, R., Athanassiou, A.: Direct transformation of edible vegetable waste into bioplastics. Macromolecules **47**, 5135–5143 (2014). https://doi.org/10.1021/ma5008557

41. Khardenavis, A.A., Suresh Kumar, M., Mudliar, S.N., Chakrabarti, T.: Biotechnological conversion of agro-industrial wastewaters into biodegradable plastic, poly β-hydroxybutyrate. Biores. Technol. **98**, 3579–3584 (2007). https://doi.org/10.1016/j.biortech.2006.11.024

42. Singh, A., Kuila, A., Adak, S., Bishai, M., Banerjee, R.: Utilization of vegetable wastes for bioenergy generation. Agric. Res. **1**, 213–222 (2012)
43. Gal, A., Brumfeld, V., Weiner, S., Addadi, L., Oron, D.: Certain biominerals in leaves function as light scatterers. Adv. Mater. **24** (2012). https://doi.org/10.1002/adma.201104548
44. Res, J.M.B., Kandari, V., Gupta, S.: Bioconversion of Vegetable and Fruit Peel Wastes in viable product. J. Microbiol. Biotechnol. Res. **2**, 308–312 (2012)
45. Kulkarni, S.O., Kanekar, P.P., Jog, J.P., Sarnaik, S.S., Nilegaonkar, S.S.: Production of copolymer, poly (hydroxybutyrate-co-hydroxyvalerate) by Halomonas campisalis MCM B-1027 using agro-wastes. Int. J. Biol. Macromol. **72**, 784–789 (2015). https://doi.org/10.1016/j.ijbiomac.2014.09.028
46. Maskow, T., Babel, W.: Calorimetrically recognized maximum yield of poly-3-hydroxybutyrate (PHB) continuously synthesized from toxic substrates. J. Biotechnol. **77**, 247–253 (2000). https://doi.org/10.1016/S0168-1656(99)00220-5
47. Witzke, D.: Introduction to properties, engineering, and prospects of polylactide polymers (1999)
48. Lucas, N., Bienaime, C., Belloy, C., Queneudec, M., Silvestre, F., Nava-Saucedo, J.E.: Polymer biodegradation: Mechanisms and estimation techniques—a review. Chemosphere **73**, 429–442 (2008). https://doi.org/10.1016/j.chemosphere.2008.06.064
49. Lunt, J., Bone, J.: Properties and dyeability of fibers and fabrics produced from polylactide (PLA) polymers. AATCC Rev. **1**, 20–23 (2001)
50. Tavanaie, M.A.: Melt recycling of poly(lactic acid) plastic wastes to produce biodegradable fibers. Polym. Plast. Technol. Eng. **53**, 742–751 (2014). https://doi.org/10.1080/03602559.2013.877931
51. Valentino, F., Morgan-Sagastume, F., Campanari, S., Villano, M., Werker, A., Majone, M.: Carbon recovery from wastewater through bioconversion into biodegradable polymers. New Biotechnol. **37**, 9–23 (2017). https://doi.org/10.1016/j.nbt.2016.05.007
52. Yan, S., Tyagi, R.D., Surampalli, R.Y.: Polyhydroxyalkanoates (PHA) production using wastewater as carbon source and activated sludge as microorganisms. Water Sci. Technol. **53**, 175–180 (2006). https://doi.org/10.2166/wst.2006.193
53. Chua, A.S.M., Takabatake, H., Satoh, H., Mino, T.: Production of polyhydroxyalkanoates (PHA) by activated sludge treating municipal wastewater: effect of pH, sludge retention time (SRT), and acetate concentration in influent. Water Res. **37**, 3602–3611 (2003). https://doi.org/10.1016/S0043-1354(03)00252-5
54. Comeau, Y., Hall, K.J., Oldham, W.K.: Determination of poly-3-hydroxybutyrate and poly-3-hydroxyvalerate in activated sludge by gas-liquid chromatography, vol 54 (1988)
55. Takabatake, H., Satoh, H., Mino, T., Matsuo, T.: PHA (polyhydroxyalkanoate) production potential of activated sludge treating wastewater (2002)
56. Bosco, F., Chiampo, F.: Production of polyhydroxyalcanoates (PHAs) using milk whey and dairy wastewater activated sludge. Production of bioplastics using dairy residues. J. Biosci. Bioeng. **109**, 418–421 (2010). https://doi.org/10.1016/j.jbiosc.2009.10.012
57. Chaleomrum, N., Chookietwattana, K., Dararat, S.: Production of PHA from Cassava starch wastewater in sequencing batch reactor treatment system. APCBEE Proc. **8**, 167–172 (2014). https://doi.org/10.1016/j.apcbee.2014.03.021
58. Bergo, P.V.A., Carvalho, R.A., Sobral, P.J.A., Dos Santos, R.M.C., Da Silva, F.B.R., Prison, J.M., Solorza-Feria, J., Habitante, A.M.Q.B.: Physical properties of edible films based on cassava starch as affected by the plasticizer concentration. Packag. Technol. Sci. **21**, 85–89 (2008). https://doi.org/10.1002/pts.781
59. Lee, S.Y.E.: Coli moves into the plastic age. Nat. Biotechnol. **15**, 17–18 (1997)
60. Yu, H., Shi, Y., Yin, J., Shen, Z., Yang, S.: Genetic strategy for solving chemical engineering problems in biochemical engineering. Proc. J. Chem. Technol. Biotechnol. **78**, 283–286 (2003)
61. Song, S., Hein, S., Steinbüchel, A.: Production of poly(4-hydroxybutyric acid) by fed-batch cultures of recombinant strains of Escherichia coli. Biotech. Lett. **21**, 193–197 (1999). https://doi.org/10.1023/A:1005451810844

62. Langenbach, S., Rehm, B.H.A., Steinbüchel, A.: Functional expression of the PHA synthase gene phaC1 from *Pseudomonas aeruginosa* in *Escherichia coli* results in poly(3-hydroxyalkanoate) synthesis. FEMS Microbiol. Lett. **150**, 303–309 (1997). https://doi.org/10.1016/S0378-1097(97)00142-0

63. Steinbüchel, A., Füchtenbusch, B.: Bacterial and other biological systems for polyester production. Trends Biotechnol. **16**, 419–427 (1998)

64. Anderson, A.J., Dawes, E.A.: Occurrence, Metabolism, metabolic role, and industrial uses of bacterial polyhydroxyalkanoates, vol. 54 (1990)

65. Lee, S.Y.: Plastic bacteria? Progress and prospects for polyhydroxyalkanoate production in bacteria. Trends Biotechnol. **14**, 431–438 (1996)

66. Reddy, C.S.K., Ghai, R., Rashmi, Kalia, V.C.: Polyhydroxyalkanoates: an overview. Bioresour. Technol. **87**, 137–146 (2003)

67. Suriyamongkol, P., Weselake, R., Narine, S., Moloney, M., Shah, S.: Biotechnological approaches for the production of polyhydroxyalkanoates in microorganisms and plants—a review. Biotechnol. Adv. **25**, 148–175 (2007)

68. Dawes, E.A.: Novel microbial polymers: an introductory overview. In: Dawes, E.A. (eds.) Novel Biodegradable Microbial Polymers, pp. 3–16. Springer, Netherlands (1990)

69. Salehizadeh, H., Van Loosdrecht, M.C.M.: Production of polyhydroxyalkanoates by mixed culture: recent trends and biotechnological importance. Biotechnol. Adv. **22**, 261–279 (2004). https://doi.org/10.1016/j.biotechadv.2003.09.003

70. Ojumu, T.V., Yu, J., Solomon, B.O.: Production of polyhydroxyalkanoates, a bacterial biodegradable polymer. Afr. J. Biotech. **3**, 18–24 (2004)

71. Haas, R., Jin, B., Zepf, F.T.: Production of poly(3-hydroxybutyrate) from waste potato starch 70503 (1C–2) communication. Biosci. Biotechnol. Biochem **72**, 70503–70504 (2008). https://doi.org/10.1271/bbb.70503

72. Solaiman, D., Ashby, R.: Biosynthesis of medium-chain-length poly(hydroxyalkanoates) from soy molasses pectin modification view project release and recovery of value-added co-products from citrus biomass by steam-explosion view project. Biotech. Lett. **28**, 157–162 (2006). https://doi.org/10.1007/s10529-005-5329-2

73. Huang, T.Y., Duan, K.J., Huang, S.Y., Chen, C.W.: Production of polyhydroxyalkanoates from inexpensive extruded rice bran and starch by *Haloferax mediterranei*. J. Ind. Microbiol. Biotechnol. **33**, 701–706 (2006). https://doi.org/10.1007/s10295-006-0098-z

74. Bhubalan, K., Lee, W.H., Loo, C.Y., Yamamoto, T., Tsuge, T., Doi, Y., Sudesh, K.: Controlled biosynthesis and characterization of poly(3-hydroxybutyrate-co-3-hydroxyvalerate-co-3-hydroxyhexanoate) from mixtures of palm kernel oil and 3HV-precursors. Polym. Degrad. Stab. **93**, 17–23 (2008). https://doi.org/10.1016/j.polymdegradstab.2007.11.004

75. Ryu, H.W., Hahn, S.K., Chang, Y.K., Chang, H.N.: Production of poly(3-hydroxybutyrate) by high cell density fed-batch culture of *Alcaligenes eutrophus* with phosphate limitation. Biotechnol. Bioeng. **55**, 28–32 (1997). https://doi.org/10.1002/(SICI)1097-0290(19970705)55:1%3c28::AID-BIT4%3e3.0.CO;2-Z

76. Cho, K.S., Ryu, H.W., Park, C.H., Goodrich, P.R.: Poly(hydroxybutyrate-co-hydroxyvalerate) from swine waste liquor by *Azotobacter vinelandii* UWD. Biotech. Lett. **19**, 7–10 (1997). https://doi.org/10.1023/A:1018342332141

77. Wu, Q., Wang, Y., Chen, G.Q.: Medical application of microbial biopolyesters polyhydroxyalkanoates. Artif. Cells Blood Substitutes Biotechnol. **37**, 1–12 (2009)

78. Eggink, G., Steinbüchel, A., Poirier, Y., Witholt, B.: International Symposium on Bacterial Polyhydroxyalkanoates (1997)

79. Atlas, R.M.: Petroleum biodegradation and oil spill bioremediation. Mar. Pollut. Bull. **31**, 178–182 (1995). https://doi.org/10.1016/0025-326X(95)00113-2

80. Dalal, J., Sarma, P.M., Lavania, M., Mandal, A.K., Lal, B.: Evaluation of bacterial strains isolated from oil-contaminated soil for production of polyhydroxyalkanoic acids (PHA). Pedobiologia **54**, 25–30 (2010). https://doi.org/10.1016/j.pedobi.2010.08.004

81. Masood, F., Hasan, F., Ahmed, S., Hameed, A.: Biosynthesis and characterization of poly (3-hydroxybutyrate-co-3- hydroxyvalerate) from *Bacillus cereus* FA11 isolated from TNT-contaminated soil. Ann. Microbiol. **62**, 1377–1384 (2012). https://doi.org/10.1007/s13213-011-0386-3

82. Hori, K., Abe, M., Unno, H.: Production of triacylglycerol and poly(3-hydroxybutyrate-co-3-hydroxyvalerate) by the toluene-degrading bacterium *Rhodococcus aetherivorans* IAR1. J. Biosci. Bioeng. **108**, 319–324 (2009). https://doi.org/10.1016/j.jbiosc.2009.04.020

83. Ni, Y.Y., Kim, D.Y., Chung, M.G., Lee, S.H., Park, H.Y., Rhee, Y.H.: Biosynthesis of medium-chain-length poly(3-hydroxyalkanoates) by volatile aromatic hydrocarbons-degrading *Pseudomonas fulva* TY16. Biores. Technol. **101**, 8485–8488 (2010). https://doi.org/10.1016/j.biortech.2010.06.033

84. Singh Saharan, B., Grewal, A., Kumar, P.: Biotechnological production of polyhydroxyalkanoates: a review on trends and latest developments. Chin. J. Biol. **2014**, 1–18 (2014). https://doi.org/10.1155/2014/802984

85. De Philippis, R., Ena, A., Guastiini, M., Sili, C., Vincenzini, M.: Factors affecting poly-β-hydroxybutyrate accumulation in cyanobacteria and in purple non-sulfur bacteria. FEMS Microbiol. Lett. **103**, 187–194 (1992). https://doi.org/10.1016/0378-1097(92)90309-C

86. Nishioka, M., Nakai, K., Miyake, M. , Asada, Y. , Taya, M.: Production of poly-β-hydroxybutyrate by thermophilic cyanobacterium, *Synechococcus* sp. MA19, under phosphate-limited conditions. Biotechnol. Lett. **23**, 1095–1099 (2001). https://doi.org/10.1023/A:1010551614648.

87. Yew, S., Jau, M., Yong, K., Abed, R., Sudesh, K.: Morphological studies of *Synechocystis* sp. UNIWG under polyhydroxyalkanoate accumulating conditions (2005)

88. Panda, B., Mallick, N.: Enhanced poly-β-hydroxybutyrate accumulation in a unicellular cyanobacterium, *Synechocystis* sp. PCC 6803. Lett. Appl. Microbiol. **44**, 194–198 (2007). https://doi.org/10.1111/j.1472-765X.2006.02048.x

89. Melnicki, M.R., Eroglu, E., Melis, A.: Changes in hydrogen production and polymer accumulation upon sulfur-deprivation in purple photosynthetic bacteria. Int. J. Hydrogen Energy **34**, 6157–6170 (2009). https://doi.org/10.1016/j.ijhydene.2009.05.115

90. Sharma, L., Mallick, N.: Enhancement of poly-β-hydroxybutyrate accumulation in Nostoc muscorum under mixotrophy, chemoheterotrophy and limitations of gas-exchange. Biotech. Lett. **27**, 59–62 (2005). https://doi.org/10.1007/s10529-004-6586-1

91. Xu, J., Wang, K., Wang, C., Hu, F., Zhang, Z., Xu, S., Wu, J.: Sodium acetate stimulates PHB biosynthesis in *Synechocystis* sp. PCC 6803. Tsinghua Sci. Technol. **20**, 435–438 (2002). https://doi.org/10.1016/S1007-0214(08)70020-8

92. Oren, A.: Microbial life at high salt concentrations: phylogenetic and metabolic diversity. Saline Syst. **4** (2008). https://doi.org/10.1186/1746-1448-4-2

93. Fernandez-Castillo, R., Rodriguez-Valera, F., Gonzalez-Ramos, J., Ruiz-Berraquero', F.: Accumulation of Poly(Q-Hydroxybutyrate) by Halobacteria, vol. 51 (1986)

94. Garcia Lillo, J., Rodriguez-Valera, F.: Effects of culture conditions on poly(β-hydroxybutyrate acid) production by *Haloferax mediterranei*. Appl. Environ. Microbiol. **56**, 2517–2521 (1990)

95. Han, J., Lu, Q., Zhou, L., Zhou, J., Xiang, H.: Molecular characterization of the phaECHm genes, required for biosynthesis of poly(3-hydroxybutyrate) in the extremely halophilic archaeon *Haloarcula marismortui*. Appl. Environ. Microbiol. **73**, 6058–6065 (2007). https://doi.org/10.1128/AEM.00953-07

96. Pramanik, A., Mitra, A., Arumugam, M., Bhattacharyya, A., Sadhukhan, S., Ray, A., Haldar, S., Mukhopadhyay, U.K., Mukherjee, J.: Utilization of vinasse for the production of polyhydroxybutyrate by *Haloarcula marismortui*. Folia Microbiol. **57**, 71–79 (2012). https://doi.org/10.1007/s12223-011-0092-3

97. Darandeh, N., Hadavi, E.: Effect of pre-harvest foliar application of citric acid and malic acid on chlorophyll content and post-harvest vase life of Lilium cv. Brunello. Front. Plant Sci. **2** (2012). https://doi.org/10.3389/fpls.2011.00106

98. Saharan, B.S., Sahu, R.K., Sharma, D.: A review on biosurfactants: fermentation, current developments and perspectives. Genet. Eng. Biotechnol. J. **2011**, 1–14 (2011)

99. Gasser, I., Müller, H., Berg, G.: Ecology and characterization of polyhydroxyalkanoate-producing microorganisms on and in plants. FEMS Microbiol. Ecol. **70**, 142–150 (2009). https://doi.org/10.1111/j.1574-6941.2009.00734.x

100. Kadouri, D., Jurkevitch, E., Okon, Y., Castro-Sowinski, S.: Ecological and agricultural significance of bacterial polyhydroxyalkanoates. Crit. Rev. Microbiol. **31**, 55–67 (2005)

101. Tunlid, A., Baird, B.H., Trexler, M.B., Olsson, S., Findlay, R.H., Odham, G., White, D.C. Determination of phospholipid ester-linked fatty acids and poly β-hydroxybutyrate for the estimation of bacterial biomass and activity in the rhizosphere of the rape plant *Brassica napus* (L.). Can. J. Microbiol. **31**, 1113–1119 (1985). https://doi.org/10.1139/m85-210

102. De Lima, T.C.S., Grisi, B.M., Bonato, M.C.M.: Bacteria isolated from a sugarcane agroecosystem: their potential production of polyhydroxyalcanoates and resistance to antibiotics. Rev. Microbiol. **30**, 214–224 (1999). https://doi.org/10.1590/s0001-37141999000300006

103. Doran-Peterson, J., Cook, D.M., Brandon, S.K.: Microbial conversion of sugars from plant biomass to lactic acid or ethanol. Plant J. **54**, 582–592 (2008)

104. Albertsson, A.C., Varma, I.K.: Recent developments in ring opening polymerization of lactones for biomedical applications. Biomacromol **4**, 1466–1486 (2003)

105. Taguchi, S., Yamada, M., Matsumoto, K., Tajima, K., Satoh, Y., Munekata, M., Ohno, K., Kohda, K., Shimamura, T., Kambe, H., et al.: A microbial factory for lactate-based polyesters using a lactate-polymerizing enzyme. Proc. Natl. Acad. Sci. U.S.A. **105**, 17323–17327 (2008). https://doi.org/10.1073/pnas.0805653105

106. Mooney, B.P.: The second green revolution? Production of plant-based biodegradable plastics. Biochem. J **418**, 219–232 (2009). https://doi.org/10.1042/BJ20081769

107. Swain, S.N., Biswal, S.M., Nanda, P.K., Nayak, P.L.: Biodegradable soy-based plastics: opportunities and challenges. J. Polym. Environ. **12**, 35–42 (2004). https://doi.org/10.1023/B:JOOE. 0000003126.14448.04

108. Nanda, P.K., Nayak, P.L., Rao, K.K.: Thermal degradation analysis of biodegradable plastics from urea-modified soy protein isolate. Polym. Plast. Technol. Eng. **46**, 207–211 (2007). https://doi.org/10.1080/03602550601152713

109. Cuq, B., Gontard, N., Guilbert, S.: Thermal properties of fish myofibrillar protein-based films as affected by moisture content. Polymer **38**, 2399–2405 (1997). https://doi.org/10.1016/S0032-3861(96)00781-1

110. Cuq, B., Gontard, N., Guilbert, S.: Proteins as agricultural polymers for packaging production. Cereal Chem. **75**, 1–9 (1998)

111. Muneer, F.: Bioplastics from Natural Polymers, pp. 1–10 (2014)

112. Keziah, V.S., Gayathri, R., Priya, V.V.: Biodegradable plastic production from corn starch. Drug Invent. Today **10**, 1315–1317 (2018)

113. Muneer, F.: Bioplastics from natural polymers (2014)

114. Abolibda, T.Z.: Physical and Chemical investigations of starch based bio-plastics, pp. 2–29. Thesis submitted for the degree of Doctor of Philosophy at the University of Leicester; University of Leicester (2015)

115. Scandola, M., Ceccorulli, G., Pizzoli, M.: Miscibility of bacterial Poly(3-hydroxybutyrate) with cellulose esters. Macromolecules **25**, 6441–6446 (1992). https://doi.org/10.1021/ma00050a009

116. Lassalle, V.L., Ferreira, M.L.: Lipase-catalyzed synthesis of polylactic acid: an overview of the experimental aspects. J. Chem. Technol. Biotechnol. **83**, 1493–1502 (2008)

117. Jafari-Sales, A.: Bioplastics and the environment. Electron.J. Biol. **13**, 274–279 (2017)

118. Fukui, T., Doi, Y.: Efficient production of polyhydroxyalkanoates from plant oils by *Alcaligenes eutrophus* and its recombinant strain. Appl. Microbiol. Biotechnol. **49**, 333–336 (1998). https://doi.org/10.1007/s002530051178

119. Eggink, G., van der Wal, H., Huijberts, G.N.M., de Waard, P.: Oleic acid as a substrate for poly-3-hydroxyalkanoate formation in *Alcaligenes eutrophus* and *Pseudomonas putida*. Ind. Crops Prod. **1**, 157–163 (1992). https://doi.org/10.1016/0926-6690(92)90014-M

120. Lageveen, R.G., Huisman, G.W., Preusting, H., Ketelaar, P., Eggink, G., Witholt, B.: Formation of polyesters by *Pseudomonas oleovorans*: effect of substrates on formation and composition of poly-(R)-3-hydroxyalkanoates and poly-(R)-3-hydroxyalkenoates. Appl. Environ. Microbiol. **54**, 2924–2932 (1988)

121. Yu, J.: Microbial production of bioplastics from renewable resources. In: Bioprocessing for Value-Added Products from Renewable Resources, pp. 585–610 (2007). ISBN 9780444521149

122. Khalid, S., Yu, L., Meng, L., Liu, H., Ali, A., Chen, L.: Poly(lactic acid)/starch composites: effect of microstructure and morphology of starch granules on performance. J. Appl. Polym. Sci. **134**, 1–12 (2017). https://doi.org/10.1002/app.45504

123. Spierling, S., Knüpffer, E., Behnsen, H., Mudersbach, M., Krieg, H., Springer, S., Albrecht, S., Herrmann, C., Endres, H.J.: Bio-based plastics—a review of environmental, social and economic impact assessments. J. Clean. Prod. **185**, 476–491 (2018). https://doi.org/10.1016/j.jclepro.2018.03.014

124. Tsou, C.H., Suen, M.C., Yao, W.H., Yeh, J.T., Wu, C.S., Tsou, C.Y., Chiu, S.H., Chen, J.C., Wang, R.Y., Lin, S.M., et al.: Preparation and characterization of bioplastic-based green renewable composites from tapioca with acetyl tributyl citrate as a plasticizer. Materials **7**, 5617–5632 (2014). https://doi.org/10.3390/ma7085617

125. Chen, G.-Q.: plastics completely synthesized by bacteria: Polyhydroxyalkanoates 17–37 (2010)

126. Brandl, H., Gross, R.A., Lenz, R.W., Fuller, R.C.: Plastics from bacteria and for bacteria: poly(beta-hydroxyalkanoates) as natural, biocompatible, and biodegradable polyesters. Adv. Biochem. Eng. Biotechnol. **41**, 77–93 (1990). https://doi.org/10.1007/bfb0010232

127. Chanprateep, S.: Current trends in biodegradable polyhydroxyalkanoates. J. Biosci. Bioeng. **110**, 621–632 (2010)

128. Davis, G.: Characterization and characteristics of degradable polymer sacks. Mater. Charact. **51**, 147–157 (2003). https://doi.org/10.1016/j.matchar.2003.10.008

129. Dornburg, V., Lewandowski, I., Patel, M.: Comparing the land requirements, energy savings, and greenhouse gas emissions reduction of biobased polymers and bioenergy: an analysis and system extension of life-cycle assessment studies. J. Ind. Ecol. **7**, 93–116 (2004). https://doi.org/10.1162/108819803323059424

130. Galindo, E., Peña, C., Núñez, C., Segura, D., Espín, G.: Molecular and bioengineering strategies to improve alginate and polydydroxyalkanoate production by *Azotobacter vinelandii*. Microbial Cell Factories **6** (2007)

131. Gironi, F., Piemonte, V.: Bioplastics and petroleum-based plastics: strengths and weaknesses. Energy Sour. Part A Recovery Utilization Environ. Effects **33**, 1949–1959 (2011). https://doi.org/10.1080/15567030903436830

132. Harding, K.G., Dennis, J.S., von Blottnitz, H., Harrison, S.T.L.: Environmental analysis of plastic production processes: comparing petroleum-based polypropylene and polyethylene with biologically-based poly-β-hydroxybutyric acid using life cycle analysis. J. Biotechnol. **130**, 57–66 (2007). https://doi.org/10.1016/j.jbiotec.2007.02.012

133. Kim, D.H.; Na, S.K.; Park, J.S.: Preparation and characterization of modified starch-based plastic film reinforced with short pulp fiber. I. Structural properties. J. Appl. Polymer Sci. **88**, 2100–2107 (2003). https://doi.org/10.1002/app.11630

134. Ibrahim, A., Ahmed, M., Suyama, T., Tokiwa, Y., Ouichanpagdee, P., Kanagawa, T., Kamagata, Y., Singh Saharan, B., Grewal, A., Kumar, P., et al.: Production of PHA from cassava starch wastewater in sequencing batch reactor treatment system. Arab. J. Chem. **2**, 1–18 (2014). https://doi.org/10.1016/j.arabjc.2015.04.008

135. Lligadas, G., Ronda, J.C., Galià, M., Cádiz, V.: Oleic and undecylenic acids as renewable feedstocks in the synthesis of polyols and polyurethanes. Polymers **2**, 440–453 (2010). https://doi.org/10.3390/polym2040440

136. Momani, B.L.: Digital WPI Interactive Qualifying Projects (All Years) Interactive Qualifying Projects Assessment of the Impacts of Bioplastics: Energy Usage, Fossil Fuel Usage, Pollution, Health Effects, Effects on the Food Supply, and Economic Effects Compared to Petr (2009)

137. Pimentel, D.: Biofuels, solar and wind as renewable energy systems: benefits and risks (2008). ISBN 9781402086533
138. Herdman, R.C.: Biopolymers: Making Materials Nature's Way. Background Paper, OTA-BP-E-102. U.S. Congress, Office of Technology Assessment (1993)
139. Vink, E.T.H., Rábago, K.R., Glassner, D.A., Gruber, P.R.: Applications of life cycle assessment to NatureWorks^{TM} polylactide (PLA) production. Polym. Degrad. Stab. **80**, 403–419 (2003). https://doi.org/10.1016/S0141-3910(02)00372-5
140. Arikan, E.B., Ozsoy, H.D.: A review: investigation of bioplastics. J. Civil Eng. Architect. **9** (2015). https://doi.org/10.17265/1934-7359/2015.02.007
141. Rujnić-Sokele, M., Pilipović, A.: Challenges and opportunities of biodegradable plastics: a mini review. Waste Manage. Res. **35**, 132–140 (2017). https://doi.org/10.1177/0734242X1 6683272
142. dos Reis, A.R., de Queiroz Barcelos, J.P., de Souza Osório, C.R.W., Santos, E.F., Lisboa, L.A.M., Santini, J.M.K., dos Santos, M.J.D., Furlani Junior, E., Campos, M., de Figueiredo, P.A.M., et al.: A glimpse into the physiological, biochemical and nutritional status of soybean plants under Ni-stress conditions. Environ. Experiment. Botany **144**, 76–87 (2017). https://doi.org/10.1016/j.envexpbot.2017.10.006

Challenges and Future of Nanotechnology

Muhammad Bilal Tahir, Muhammad Sagir, and Muhammad Rafique

Nanotechnology has very vast and broaden view based on realism and cross-fertilization which yield new materials in science such as nanostructured materials, nano-composites, and nano-hybrids. [1]. The advancement in technological sector is increasing day by day due to advancement in nanotechnology. It is expected that utilization of nanostructures and nanomaterials will be increased rapidly in the coming years. There are numerous difficulties in the field of nanotechnology, yet a couple of them are at higher priority than others. The nanomaterials are very important due to their significant properties and variety of applications, but, if they are mishandled, then it became destructive and tremendously injurious. Therefore, it is essential to advance the technology which can measure the mishandled nanomaterials quantitatively in environment. Some nanomaterials are very toxic for humans and environment. Therefore, researchers should focus on new techniques to standardize the measure of hazardous effect of nanomaterials on human health.

Nanomaterials have been widely utilized in multiple applications since the last few as described below. Several nanomaterials have been coated on the glass windows for the prevention of dust such as titanium nanoparticles. Nanoparticles are used in sunglasses as anti-reflection coating and anti-scratch coatings. Nanoparticles are implemented in textile to make waterproof, stain-resistant or wrinkle-free fibers. Nanofibers are used in ski jacket to make it waterproof and window-proof. In

M. B. Tahir
Department of Physics, Khwaja Fareed University of Engineering and Information Technology,
Rahim Yar Khan, Pakistan
e-mail: m.bilaltahir7@gmail.com

M. Sagir (✉)
Department of Chemical Engineering, Khwaja Fareed University of Engineering and Information
Technology, Rahim Yar Khan, Pakistan
e-mail: dr.msagir@kfueit.edu.pk

M. Rafique
Department of Physics, University of Sahiwal, Sahiwal 57000, Pakistan

the future, nanotechnology could be able to introduce smart clothes and wearable electronic clothes with sensors and Internet connection.

Nanomaterials have also been employed in sports such as nanotubes in tennis rackets to enhance their resistance and torsion. Moreover, nanoparticles such a titanium oxide are also used in sunscreens to make them transparent. Recently, L'Oreal discovered that moisturizers become transparent when grounded to 60 nm.

In the future, televisions, with carbon nanotubes and field-effect display, are expected to be thinner with better quality as compared to traditional televisions [2]. Moreover, nanotechnology is expected to improve the storage capacity and processing speed in electronics and communication devices by the implementation of quantum dots and nanolayers. This implementation will revolutionize the way of communication. It is predicted that nanoparticles as catalysts will improve environment by improving combustion efficiency of plants and vehicles. New smart materials will be used for vacuum and as lubricants and seals. Nanotechnology will change the medical industry by introducing new and innovative drug delivery mechanism. The things, we see in movies, will then actually be acquired by the utilization of nanotechnology. Engineering of materials will be revolutionized by new nano-hybrid techniques and tools. It is anticipated that the energy crisis can be solved with cleaner energy sources such as quantum well solar, the outer space can be understood, in a better way, through the utilization of lightweight vehicles and very small robotic systems. Climate change issues can be resolved by utilizing the membranes that can filter dust or salt from water and pollutants from air. Nanotechnology can also improve the security protocols by implementing new materials and security devices.

Nanotechnology has been employed health sector for drug delivery. The nanoparticles can enter the human body via skin or inhalation and can cause several hazardous effects on human health. It has been addressed that nanoparticles can easily enter the human body as compared to bulk particles. A limited data about size and exposure time of nanoparticles is available nowadays. Therefore, it is necessary to gather and analyze more data on these features of nanoparticles in future. As concluding remarks, it is necessary to discover new nanomaterials and to investigate their effect on environment and different species for a bright future.

References

1. Gopakumar, G., Menon, H., Ashok, A., Nair, S.V., Shanmugam, M.: Two dimensional layered electron transport bridges in mesoscopic TiO_2 for dye sensitized solar cell applications. Electrochemica Acta **267**, 63–70 (2018)
2. Kanellos M (2005) Carbon TVs to edge out liquid crystal and plasma. News.com. [Online Document], Jan 2005

Printed in the United States
by Baker & Taylor Publisher Services